DIABETES
Its Physiological and Biochemical Basis

DIABETES

Its Physiological and Biochemical Basis

Edited by
J. Vallance-Owen
Professor of Medicine, The Queen's University of Belfast

MTP

Published in UK by
MTP Press Ltd
St Leonard's House,
Lancaster

ISBN 0 85200123-1
Printed in Great Britain
by Blackburn Times Press, Blackburn

Contents

List of Contributors

Foreword

1. Proinsulin and insulin biosynthesis   1
   Arthur H. Rubenstein, David L. Horwitz and Donald F. Steiner

2. The pancreas and insulin release   31
   S. J. H. Ashcroft and P. J. Randle

3. The gastrointestinal tract and diabetes mellitus   63
   K. D. Buchanan

4. The liver in glucose homeostasis in normal man and in diabetes   93
   Philip Felig

5. Atherosclerosis and disorders of lipid metabolism in diabetes   125
   R. W. Stout, E. L. Bierman and J. D. Brunzell

6. Synalbumin insulin antagonism   171
   J. Vallance-Owen and J. S. Bajaj

Index   203

List of Contributors

S. J. H. Ashcroft
Lecturer, Department of Clinical Biochemistry, Radcliffe Infirmary, Oxford, UK

J. S. Bajaj
Professor, All-India Institute of Medical Sciences, New Delhi, India

Edwin L. Bierman
Professor and Head, Division of Metabolism, Endocrinology and Gerontology, Department of Medicine, University of Washington School of Medicine, Seattle, USA

John D. Brunzell
Associate Professor and Chief, Lipid Metabolism Section, Division of Metabolism, Endocrinology and Gerontology, Department of Medicine, University of Washington School of Medicine, Seattle, USA

K. D. Buchanan
Senior Lecturer, Department of Medicine, Queen's University of Belfast, Northern Ireland

Philip Felig
Professor and Vice Chairman, Department of Internal Medicine, Yale University School of Medicine, New Haven, Connecticut, USA

David L. Horwitz
Assistant Professor, Department of Medicine, Pritzker School of Medicine, University of Chicago, USA

P. J. Randle
Professor, Department of Clinical Biochemistry, Radcliffe Infirmary, Oxford, UK

Arthur H. Rubenstein
Professor, Department of Medicine, Pritzker School of Medicine, University of Chicago, USA

Donald F. Steiner
Professor, Department of Biochemistry, Pritzker School of Medicine, University of Chicago, USA

R. W. Stout
Senior Lecturer, Department of Medicine, Queen's University of Belfast, Northern Ireland

J. Vallance-Owen
Professor, Department of Medicine, Queen's University of Belfast, Northern Ireland

Foreword

Order . . . chaos . . . new order this is the usual pattern of advancing knowledge in any problem until the correct solution is eventually found. At present, considerable chaos seems to surround the condition of diabetes mellitus which continues to be defined as a chronic disorder of carbohydrate and fat metabolism characterised by hyperglycaemia and glycosuria. There are some clinicians and investigators who are prepared to go further and say categorically that it depends on a deficiency of insulin resulting from either a deficiency of its supply or a diminution in its effects. Today, diabetes covers the widest clinical range of any speciality with dermatological, metabolic, neurological, ophthalmic and psychiatric as well as renal and vascular problems. Moreover, as more and more medical conditions are found to be associated with abnormalities of carbohydrate and lipid metabolism, it is not surprising that more and more interest is being taken in this condition of diabetes mellitus. Also, it is clear that many more people are constituted as diabetics than was previously realised.

Nevertheless, in spite of, or perhaps because of, its obvious need and interest, the study of diabetes and, in particular the study of its aetiology, is complex, complicated and continues to cause considerable controversy. This present book has been written because of the need to bring together work that is going on at the growing points of diabetic medical research and by people, who have been for many years, and who still are, actively engaged in this field. The cause of diabetes is not yet known but it may ultimately be found to be associated with insulin formation; the insulin release mechanism from the pancreas; a failure of the pancreas to respond to various stimuli including other hormones, notably from the gut; possibly a derangement of liver function which is so central to the understanding of carbohydrate and fat metabolism, or to an anti-insulin factor acting peripherally. This volume of recent information with its several contributors affords the reader authoritative material on a number of subjects pertaining to the aetiology and to the better understanding of the whole syndrome of diabetes mellitus.

Behind any angulation or imperfection in these remarks there does rest with me the profound conviction that the responsibilities of the researchers in medicine and in diabetes, in particular, have never been heavier than now nor their opportunities greater; each one of us who is privileged to be in this position must bear constantly in mind that it is his individual duty to set such an example of research practice, to advance knowledge as best he can, and to so train doctors and biochemists that there will always be among our pupils those whom we have inspired and fitted out to bring ideals of original research nearer to realisation than we have been able to do ourselves.

J. Vallance-Owen

1
PROINSULIN AND INSULIN BIOSYNTHESIS

Arthur H. Rubenstein, David L. Horwitz and Donald F. Steiner

INTRODUCTION

There is increasing evidence to suggest that diabetes mellitus may represent a heterogenous group of disorders characterised by glucose intolerance. Defects in the number or quality of insulin receptors' or in their coupling to messengers which effect insulin's biological activity in tissues which are sensitive to this hormone, may possibly underlie some diabetic syndromes. In other cases, circulating insulin antagonists and excessive levels of hormones which oppose insulin's action may be aetiologically involved in the precipitation of the diabetic state. Nevertheless, a large body of evidence suggests that the defect may be confined largely, if not entirely, to the islet, or β-cell organ. At present, little precise information is available regarding the origin and differentiation of islet tissue, or the regulation of the total β-cell mass. On the other hand, considerable progress has been made in understanding the functional activities of the β-cell and in examining these for abnormalities which might be causally related to diabetes. In this chapter we will concentrate on the biosynthetic pathway of insulin and those factors which may be important in regulating this process. In addition, a number of advances in our understanding of islet cell function which have arisen as a consequence of the discovery of proinsulin will be discussed.

INSULIN BIOSYNTHESIS

Prior to 1967, the view was widely held that insulin synthesis *in vivo* was

1

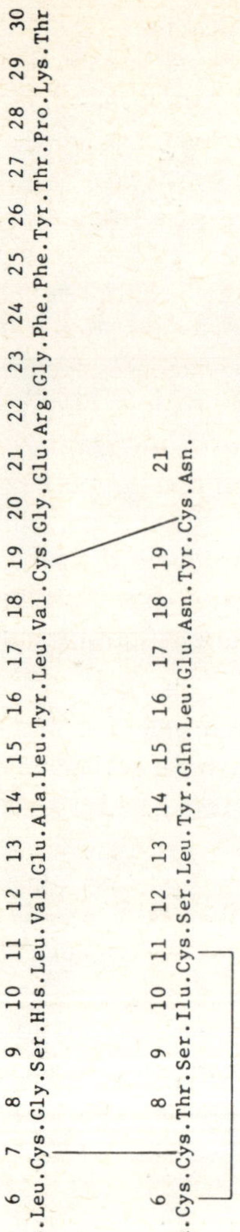

1   2   3   4   5   6   7   8   9   10 11 12 13 14 15 16 17 18 19 20 21 22 23 24 25 26 27 28 29 30
Phe.Val.Asn.Gln.His.Leu.Cys.Gly.Ser.His.Leu.Val.Glu.Ala.Leu.Tyr.Leu.Val.Cys.Gly.Glu.Arg.Gly.Phe.Phe.Tyr.Thr.Pro.Lys.Thr

1   2   3   4   5   6   7   8   9   10 11 12 13 14 15 16 17 18 19 20 21
Gly.Ilu.Val.Glu.Gln.Cys.Cys.Thr.Ser.Ilu.Cys.Ser.Leu.Tyr.Gln.Leu.Glu.Asn.Tyr.Cys.Asn.

Figure 1.1 Structure of human insulin.

accomplished by combination of the separately synthesised A and B chains. The structure of human insulin, showing these chains linked by two disulfide bridges is shown in Figure 1.1. Although inactive zymogen forms of a variety of extracellular enzymes, most notably proteases, had been known for many years, the discovery that insulin was synthesised by way of a single chain precursor, named proinsulin[1], suggested that limited proteolysis might play a more general role in protein biosynthesis. In the past decade precursor forms of a variety of small peptide hormones as well as of many viral capsule proteins[2 3], serum albumin[4 5] and even connective tissue structural proteins such as collagen[6] have been identified. A distinctive feature of insulin biosynthesis which sets it apart from classical zymogen activation is the intracellular proteolytic conversion of the precursor to the hormone prior to its storage and secretion from the β-cells[7]. The intracellular localisation and precise mechanism of this process in the β-cell is an important problem for the cell biologist as well as for the endocrinologist. The synthetic mechanism also may provide a useful model for the study of the biosynthesis of such important cellular constituents as membrane-localised proteins, various intracellular organelles, and perhaps even of some cellular enzymes[8].

Structure and properties of proinsulin

Proinsulin consists of a single polypeptide chain ranging in length from 78 (dog) to 86 (human, horse, rat) amino acid residues[9 10]. The variations in

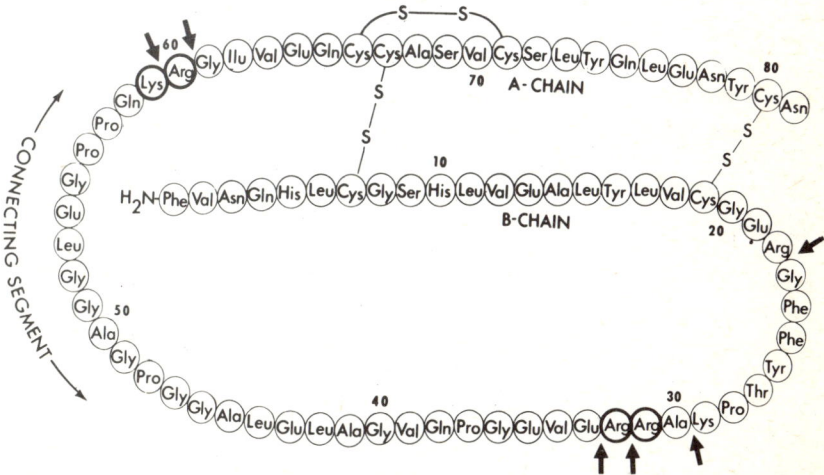

Figure 1.2 Covalent structure of bovine proinsulin. Arrows indicate sites of tryptic cleavage. (Reproduced in modified form from Nolan, *et al.*, 1971).

length in the mammalian proteins occur only in the connecting polypeptide portion which links the carboxy terminus of the insulin B chain to the amino terminus of the insulin A chain. The primary structure of bovine proinsulin[11][12] is shown in Figure 1.2. The known mammalian proinsulins have pairs of basic residues at either end of the connecting peptide which link the connecting polypeptide to the insulin chains. These residues are excised during the conversion of proinsulin to insulin, giving rise to native insulin plus the remainder of the connecting polypeptide segment lacking amino- or carboxy- terminal basic residues. This peptide has been designated the C-peptide.

Despite its considerably larger size, proinsulin is closely similar to insulin in many properties, including solubility, isoelectric point[8], self-associative properties[13] and reactivity with insulin antisera[14-16]. These observations, and a variety of preliminary evidence from other studies, strongly suggest that the conformation of the insulin moiety in proinsulin is nearly identical to that of insulin itself[8]. It is noteworthy that the connecting peptide is much larger than would seem to be required to bridge the short 8 Å gap between the ends of the B and A chains. The significance of this descrepancy is not yet understood. The connecting peptide is probably folded over a portion of the surface of the insulin molecule, but it does not completely mask the 'active site', since intact proinsulin exhibits 3–5% biological activity in several systems *in vitro*[17][18]. It is unlikely that any significant cleavage or 'activation' of proinsulin occurs in these tissues to account for this low intrinsic activity[19]. Similarly low but definite levels of biological activity have been observed in the case of several other peptide hormone precursors. The connecting peptide does not appear to obscure those monomer surfaces which interact in the formation of dimers and even of hexamers[20][20a]. A hypothetical arrangement of the connecting peptide moiety in a proinsulin hexamer is shown in Figure 1.3. This kind of hexameric structure, with the C-peptide oriented externally, could possibly facilitate the efficient conversion of proinsulin to insulin in the β-cells. The three-dimensional structure of proinsulin has not yet been determined, although crystallisation in several forms has been accomplished by Low and co-workers[21].

The precursor relationship of proinsulin to insulin

By means of isolated islet preparations[22][23] the precursor–product relationship between proinsulin and insulin have been carefully documented in a variety of biosynthetic experiments[24-26]. As mentioned above, the proteolytic conversion of proinsulin to insulin is a strictly intracellular

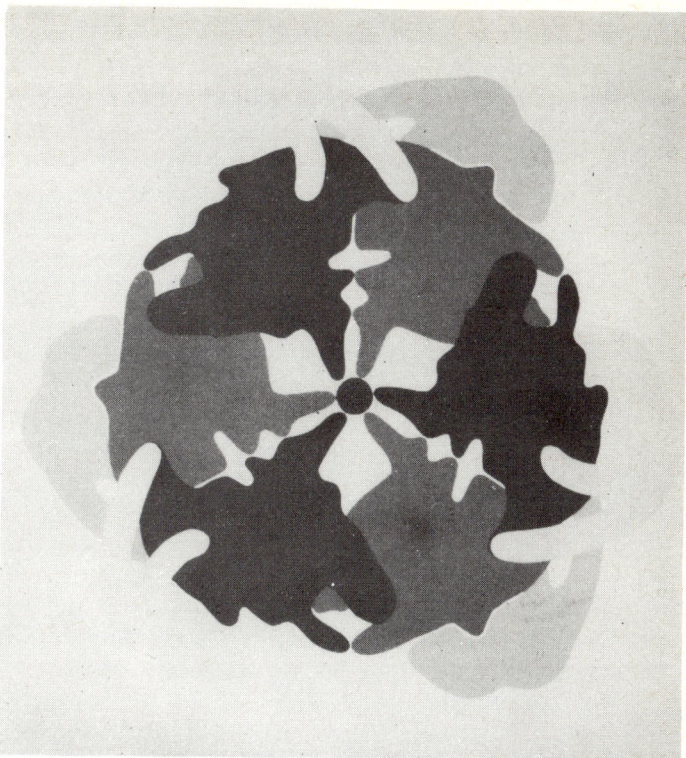

Figure 1.3 Hypothetical proinsulin hexamer arrangement as viewed along the three-fold **axis** of the hexamer. The connecting peptide portion is shown in lighter grey around the periphery of the darker outline of an insulin hexamer arranged according to the data of Blundell, *et al.,* (1971). The central density represents the two zinc atoms in coordination linkage at the six (3 above and 3 below hexamer plane) histidine side chains at position 10 in the B chain.

process[24], and it is not inhibited by cycloheximide or other inhibitors of protein synthesis, indicating that continuous protein synthesis is not necessary for the transformation of proinsulin to insulin. By means of polyacrylamide gel electrophoresis, Clark and Steiner[27] have shown the existence in rats of two proinsulins, one corresponding to each of the two insulins. Both rats and mice[28] have two non-allelic insulins, each arising from a separate proinsulin gene, presumably due to duplication of this gene during the evolution of the rodentia. The two rat insulins as well as their corresponding C-peptides differ at two positions each[29] [30]; the proinsulins thus differ at four positions. In addition to the intact proinsulins, several intermediate forms have been identified in rat islets[27] [32]. These partly

5

cleaved forms of rat proinsulin have two chains, but still retain the connecting peptide, or fragments thereof, attached to the A or B chains. During a 'chase' incubation of islets labelled with ^{3}H-leucine, radioactivity is transferred from these components, as well as from proinsulin, to insulin[27].

Comparative studies of insulin biosynthesis in the cod[33], the anglerfish[34], and in primitive vertebrates such as cyclostomes[35], indicate the formation and cleavage of a proinsulin similar in size to the mammalian proteins. A requirement for trypsin-like cleavage has been demonstrated for both of the fish proinsulins, and an interesting intermediate cleavage form, having an N-terminal tripeptide A-chain extension, has been isolated from anglerfish islets by Yamaji *et al.*[36]. A number of reports have appeared of the biosynthesis, isolation and characterisation of intermediate forms of mammalian proinsulins in various species[12 32 37 38].

The biosynthetic organisation of the β-cell

The β-cells of the islet of Langerhans share many features with other secretory cells (Figure 1.4). The participation of the Golgi apparatus in the formation of β-granules was suggested as early as 1944 by Hard[39] and later confirmed by Munger[40] by electron microscopy. He identified granules with altered morphology near the Golgi body and called these 'progranules'. Subsequent studies have demonstrated that newly synthesised peptide material passes via the Golgi apparatus into the β-cell secretory granules in a temporal sequence that is strikingly similar to that occurring in the pancreatic acinar cells[41].

It is now well established that proinsulin, in common with many other exportable proteins, is synthesised by ribosomes associated with the rough endoplasmic reticulum and it comprises the major biosynthetic product found in the microsomal fraction[42]. If a precursor form larger than proinsulin is the initial synthetic product, or for example, if the mRNA for proinsulin contains several copies of the proinsulin gene in a tandem arrangement so that a larger peptide containing several proinsulin transcripts is synthesised, one might expect to see a significant delay in the appearance of proinsulin when peptide labelling is initiated. Since this does not occur we must conclude that if higher molecular weight precursors indeed exist, these must be converted very rapidly to the 9000-dalton proinsulin.

Unfortunately, it has not yet been possible to isolate the proinsulin messenger RNA. There is no good agreement on the size of the islet polysomes actively engaged in the synthesis of proinsulin. One recent study

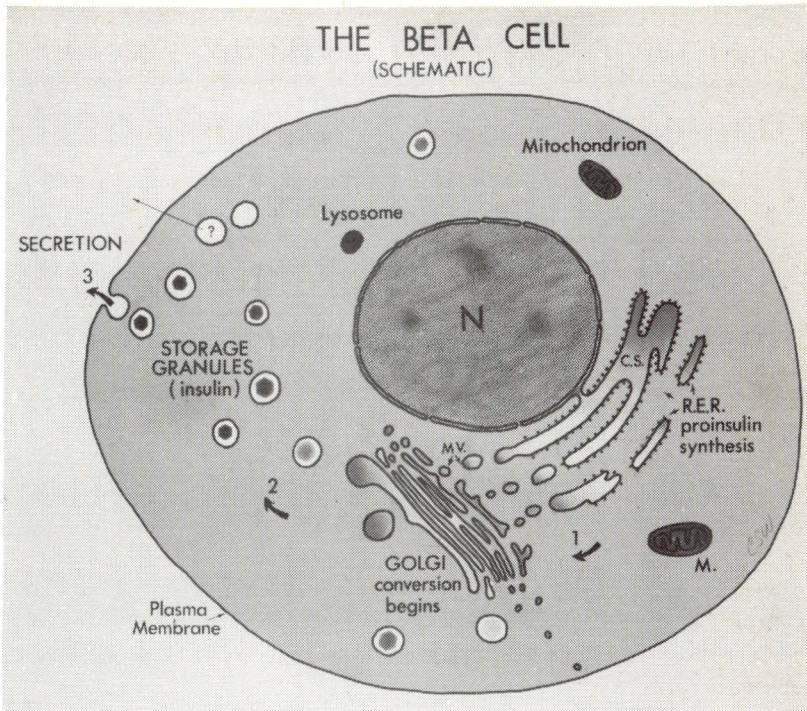

Figure 1.4 Schematic summary of the insulin biosynthetic machinery of the pancreatic beta cells. (R.E.R. = rough endoplasmic reticulum; M. V. = micro-vesicles).

concludes that in rat islets these are mainly trisomes, a size consistent with the expected mRNA length of about 258 nucleotides required to encode the 86 residue proinsulin polypeptide[43], while a more recent report suggests a somewhat larger translational unit size (pentasomes) in fetal calf pancreas[44]. These reports reflect current uncertainties as to the size of the putative mRNA as well as of the initial polypeptide synthetic product. However, no convincing indications of larger precursor forms have been found, aside from evidence with fetal calf pancreas slices suggestive of the expected role of an N-terminal residue of methionine in the initiation of proinsulin biosynthesis[45].

After biosynthesis, peptide chain folding, and sulphydril oxidation to the native structure have occurred, newly synthesised proinsulin is transported by an energy requiring process[46][47] to the Golgi apparatus. Peak labelling in the Golgi apparatus occurs at about 30 minutes after labelling the cells, with relatively little radioactivity remaining in this region after an hour[48][49].

However, the half time for conversion of proinsulin to insulin in intact rat islet cells is about one hour[7]. These observations suggest that while conversion is probably initiated in the Golgi apparatus or in the progranules as these leave the Golgi region, it continues for many additional hours within the secretory granules as they collect and mature in the cytosol.

Conversion of proinsulin to insulin

Although the proteolytic enzymes responsible for the conversion of proinsulin to insulin have not been definitely identified, several lines of evidence give some indications as to their nature. These include: (1) the known structures of the cleavage products and of a number of intermediate cleavage forms, (2) model studies with known proteolytic enzymes, and (3) the detection of converting enzyme activities in whole islet preparations or appropriate subcellular fractions[50]. The major types of proteolytic cleavage required for the conversion of proinsulin to insulin are shown in Figure 1.5. This scheme envisions the combination of a trypsin-like protease with another having specificity similar to that of carboxypeptidase B. The latter enzyme is necessary to remove the C-terminal basic residues left behind after tryptic cleavage, giving rise to the important naturally-occurring products, native insulin and the C-peptide. We have shown that appropriate mixtures of pancreatic trypsin and carboxypeptidase B can quantitatively convert proinsulin to insulin *in vitro*[51]. This model system can account for the known major intermediate forms and products that occur naturally in pancreatic extracts. In some species, such as rats, additional cleavages occur in the C-peptide region of proinsulin which appear to be due to a protease having chymotrypsin-like activity[32][38]. The role of this additional C-peptide cleavage in conversion remains unclear, however, and it probably occurs only in those species such as pigs, rats and humans where the proinsulin C-peptide contains a region of high chymotryptic sensitivity in its amino acid sequence. The occurrence of these cleavages suggests that the β-granules may contain a mixture of proteases, many of which are similar to those that occur in the exocrine pancreas. Thus the cleavage of precursor forms at specific sites may be dictated both by the high sensitivity of certain regions in the substrate molecules to a variety of known proteases, as well as by restricted specificities or special substrate adaptations on the part of the converting proteases.

Thus far we have had only partial success in demonstrating the existence of trypsin-like and carboxypeptidase B-like proteolytic activities in isolated β-cell granule preparations[37][50]. However, crude granule fractions from

PROBABLE ROUTE OF CONVERSION OF PROINSULIN TO INSULIN
(E₁ RATE-LIMITING)

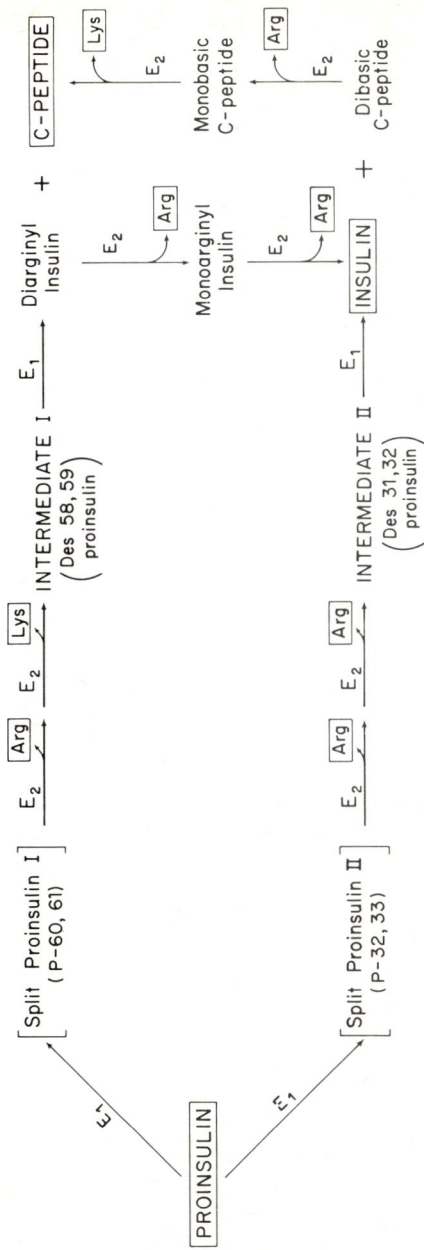

Figure 1.5 Stages in the cleavage of proinsulin by the combined action of trypsin-like and carboxypeptidase B-like proteases. (See text for further discussion of this model system.)

E_1 – Endopeptidase (Trypsin-like)
E_2 – Exopeptidase (CPase B-like)

rat islets previously labelled with ^{3}H-leucine convert the endogenously labelled proinsulin to insulin during incubation at 37°C and pH 6.3. The initial rates of conversion are comparable to those observed *in vivo*[52]. Externally added labelled bovine or rat proinsulin is not converted, however, indicating that the proteolysis is occurring only within the particulate elements. Disruption of the granules by extremes of pH, various detergents, sonication or repeated freeze-thawing markedly inhibits the conversion[52]. However, Zühlke and co-workers[52a] have recently shown that disrupted secretion granule fractions convert proinsulin to insulin-like components in a DFP-sensitive reaction when incubated at pH 6.5 with considerably higher concentrations of proinsulin than were used in our earlier experiments[37]. These workers also have described the splitting of a synthetic heptapeptide derived from the amino acid sequence of porcine proinsulin (i.e. residues 28–34 having ^{14}C–arginine at position 31) by islet homogenates and crude secretion granule fractions (Zühkle and co-workers, unpublished). Free ^{14}C-arginine could be detected, again providing evidence for the action of both an endopeptidase and a carboxypeptidase B-like enzyme.

Sun, Lin and Haist[53] have provided evidence that the proinsulin converting activity of disrupted granule fractions may be localised in a membrane-containing fraction. Such a membrane association of the endopeptidase activity might limit proteolytic activity *in vivo* and thus afford a means for protecting insulin from further proteolytic degradation in the granules. The possiblity that the converting enzymes are related to some of the lysosomal proteases has been considered by several workers[8]. However, little of the currently available information on these enzymes is consistent with a lysosomal origin. Clearly, any convincing identification of converting enzyme activity in islets will necessarily include evidence for a granule and/or Golgi localisation and proof that all of the known natural products and intermediate proteins and not just insulin, are produced by the enzyme(s).

β-Granule formation

Morphological studies of newly formed secretory granules in a variety of cells suggest that these particles undergo biochemical maturation after their formation in the Golgi apparatus. Thus, in the β-cells the 'progranules' characteristically are less dense than the mature granule inclusions and have a uniform density throughout[40]. A variety of biochemical changes undoubtedly takes place in these granules as they remain in the cytosol, including the proteolysis of proinsulin to insulin. Morphological studies of mature insulin secretion granules indicate that the dense central inclusion consists mainly of crystalline insulin[54]. Thus, as insulin is liberated from proinsulin it evidently tends to crystallise with zinc. The C-peptide liberated in the conversion process probably remains in the clear space surrounding the dense insulin crystals since there is no evidence for co-crystallisation of the C-peptide with insulin. Since proinsulin strongly retards insulin crystallisation *in vitro*[20a], it is likely that conversion must first proceed largely towards completion before insulin crystallisation can occur in the granules.

The role of zinc in secretion granule formation is an intriguing problem. The available evidence indicates that most of the islet zinc is present in the granules and is liberated proportionately with insulin during secretion[55][56]. How zinc is accumulated within the granules is not known, but proinsulin and insulin both have the ability to bind zinc[13]. However, the insulins of a few species, including the guinea pig and coypu[57], and the hagfish[58], lack the histidine residue at position 10 of the B chain required for zinc binding during the association of insulin dimers into hexamers. Mammalian proinsulins can bind larger amounts of zinc than insulin without precipitating from solution[59]. Thus proinsulin could conceivably play a role in zinc accumulation in the islet cells. These bound metal ions might then influence the conversion process and subsequently aid in the sequestration of the newly-formed insulin as osmotically inactive and biochemically stable crystalline inclusions.

It is also possible that the creation of appropriately acid conditions for insulin crystallisation to occur in the granules may be coupled biochemically to the conversion process. Optimum conditions for proinsulin folding and sulphydril oxidation are presumably somewhat above neutrality. But, as the new secretory products move to the Golgi apparatus and proteolytic conversion begins, the cationic arginine and lysine residues released in this process may diffuse from the granules and be replaced by hydrogen ions resulting in a gradual decrease in the intragranular pH. This gradual acidification as the granules mature could create appropriate conditions for the formation of the crystalline zinc inclusions. These conditions all underscore the close topographical and biochemical integration of the biosynthesis of proinsulin, its intracellular transport, proteolysis, and storage as insulin in the β-cells.

REGULATION OF INSULIN BIOSYNTHESIS

Just as glucose is the pre-eminent stimulus for insulin secretion, it also is of major importance as a regulator of insulin biosynthesis. Although the mechanism by which glucose stimulates insulin biosynthesis is not fully understood, it seems likely that some kind of signal generated either by glucose or one of its metabolites in the β-cells serves to stimulate both insulin secretion and biosynthesis. Thus mannoheptulose inhibits both responses to glucose, while glyceraldehyde stimulates both processes[60][61]. The glucose signal evidently is still effectively generated in a low calcium medium even though the extrusion of granules is inhibited under these conditions[8][62]. Similarly in fetal pancreas glucose strongly stimulates insulin biosynthesis[63][64] but not secretion, indicating that glucose responsivity is present even though secretory mechanisms are not yet fully competent. Sulphonylurea compounds, on the other hand, appear to act directly on some aspect of granule extrusion, since they only cause insulin secretion without significant stimulation of biosynthesis[8]. All these observations indicate that insulin biosynthesis is regulated directly by signals generated by glucose rather than by feed-back loops operating to sense the size of the granule population in the β-cells. The apparent lack of such feed-back mechanisms may explain the frequent observation of granule autophagy under conditions where insulin secretion is blocked either pharmacologically[65] or congenitally[66].

It is noteworthy that the stimulatory effect of glucose on protein synthesis in the β-cell is highly selective for proinsulin biosynthesis: the rates of translation of other cellular proteins are enhanced to a much lesser extent[67][68]. The glucose effect is very rapid and does not depend on the synthesis of new messenger RNA; it rather appears to be due to selective translational enhancement. Thus actinomycin D initially does not inhibit the stimulation of biosynthesis due to glucose, although it does appear to inhibit a subsequent phase of further enhancement of the biosynthetic rate beginning about 40 minutes after glucose stimulation[8][67][68]. It is likely that this further increase in synthesis is due to the enhancement of both messenger RNA transcription as well as the formation of additional ribosomes on which proinsulin mRNA molecules can be translated. In keeping with these possibilities, glucose has been reported to increase the synthesis of RNA in isolated islets of Langerhans[67]. Presumably all these highly anabolic effects of glucose lead also to the stimulation of β-cell proliferation as has recently been described by Anderson[68a]. Thus islet hyperplasia as well as unusually high rates of insulin biosynthesis and secretion normally occur in many hyperglycaemic states. Clearly any inherited or acquired deficiency in glucose responsiveness in the β-cells may have far-reaching consequences in terms of the size and functional competence of the islet organ.

C-peptide structure and properties

Due to the localization of the conversion process within secretion granules, the C-peptide accumulates along with insulin in equimolar amounts[69] and is secreted along with the hormone by exocytosis of the granule contents[70]. The amino acid sequences of C-peptides from a number of species have been combined in the composite diagram shown in Figure 1.6[29 31 71 73] These peptides exhibit a much higher rate of mutation acceptance than do the corresponding insulins, a finding consistent with the possibility that this

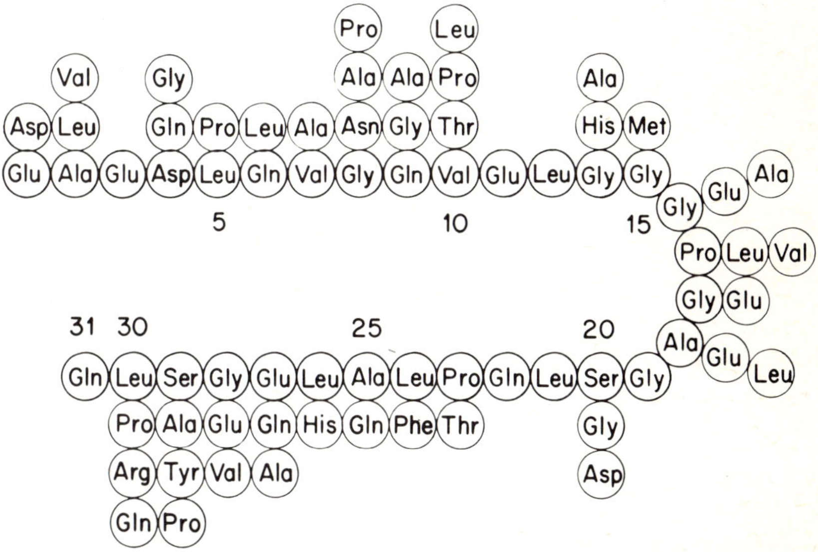

Figure 1.6 Amino acid sequence of human proinsulin C-peptide combined with the known substitutions occurring in eight other mammalian and one avian C-peptide shown alongside. Deletions occur in the dog (residues 4-11), pig (residues 18 and 19), sheep and ox (residues 22-26), and guinea pig (residues 25 and 26). (These sequences do not include the basic residues at either end which link the C-peptide to the insulin chains).

region of proinsulin does not contain an active centre for a specific hormonal function. Among known proteins, only the fibrinopeptides have a higher rate of mutation acceptance than the proinsulin C-peptides. The much lower rate of mutation acceptance of insulin, however, is similar to that of many other functional proteins such as haemoglobin or cytochrome C [74]. Certain regions of the relatively large connecting peptide may serve specific functions, such as facilitating the folding of the proinsulin polypeptide chain and the formation of the correct disulphide bonds[75] or guiding the enzymatic cleavage of proinsulin to insulin. Several acidic residues are consistently present in the connecting peptides. These tend to offset the cationic charges due to the basic residues at the cleavage sites so that the isoelectric pH of proinsulin is nearly the same as that of insulin, i.e. pH 5.1–5.5 [8]. It is also possible that translational constraints such as mRNA sequence, size or secondary structure may play a role in dictating the primary structure of some regions within proinsulin.

ISLET SECRETORY PRODUCTS

Serum proinsulin

Immunoreactive material similar in molecular weight to proinsulin[76 77] has been found in normal human plasma and urine samples. Although it would be desirable to measure human proinsulin and its intermediate fractions (proinsulin-like component, PLC) by direct immunoassay in unextracted serum, this has not been possible. The reasons lie in the cross-reactivity of proinsulin with insulin, on the one hand, and the C-peptide on the other. As all three of these peptides have been identified in the circulation, a preliminary step is required to separate them from each other. The most commonly used approach has involved gel filtration of serum followed by measurement of the column fractions in the insulin immunoassay. In order to calculate the absolute level of proinsulin and insulin, fractions of the earlier eluting peak are read from a human proinsulin standard, while those comprising the second peak are measured against a human insulin standard. Most investigators have expressed the values of proinsulin in terms of the insulin standard (Figure 1.7), because human proinsulin is not readily available.

Another method for separating insulin from proinsulin has been described by Kitabchi and his co-workers[78]. These investigators have used an enzyme which is relatively specific for the proteolytic degradation of insulin, but not proinsulin. Measuring samples in an insulin assay before and after incubation with this enzyme (insulin specific protease, ISP) should enable one to determine the relative concentrations of the two peptides. The accuracy of this method is limited, however, especially at low serum immunoreactive insulin (IRI) concentrations, because the degradation of insulin is generally incomplete. This may be due, in part, to the presence of non-competitive inhibitors of the enzyme in plasma[79 80].

Characterisation of serum proinsulin

In most studies PLC has been measured after gel filtration of sera on Sephadex or Bio-Gel columns equilibrated in acetic acid, borate or veronal buffers. Because of its higher molecular weight, PLC elutes before insulin

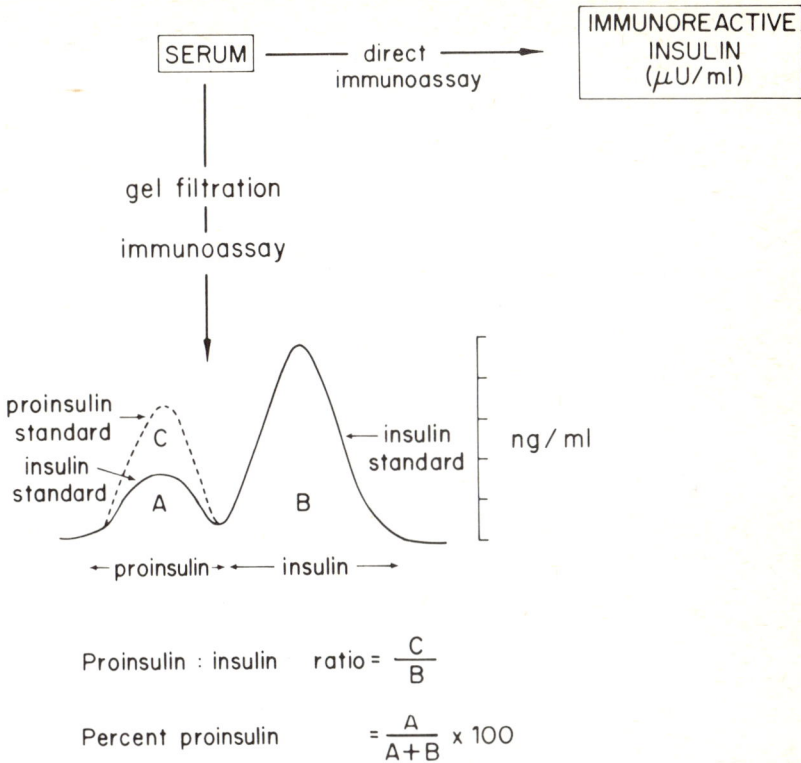

Figure 1.7 A schematic diagram illustrating the measurement of proinsulin and insulin in serum. Serum is measured directly in the immunoassay (immunoreactive insulin–IRI). It is then gel filtered to separate the proinsulin and insulin and each fraction is assayed in the insulin radioimmunoassay. The early eluting peak (proinsulin-like-component) is read from the human proinsulin standard curve (curve C), or alternatively from the human insulin standard (curve A). The sum of the individual fractions in each peak (corrected for volume) is the proinsulin and insulin concentration, respectively.

and may be identified in either the insulin or human C-peptide immunoassays. However, these methods do not differentiate the two chain proinsulin intermediates[11] from the single chain precursor and additional techniques are required to demonstrate their presence in the circulation.

Lazarus et al.[81] have described the presence of a proinsulin intermediate, in addition to proinsulin, in the serum of a patient with a surgically-documented carcinoma of the pancreatic islets. In addition to this intermediate form, other components which react in the insulin assay have been identified in sera of patients with islet cell tumours. Thus Gorden et al.[82] have described a proinsulin-like component which eluted ahead of the proinsulin marker on a 1.5 × 90 cm column of Sephadex G-50 in a subject with an islet cell carcinoma. Yet another form of immunoreactive insulin, which has been named 'big, big insulin' has been described in the plasma of an insulinoma suspect by Yalow and Berson[83]. This component has a molecular weight greater than 100 000, is immunochemically identical to human insulin, is more basic than porcine or human insulin and is rapidly transformed by trypsin to an insulin-like component. In several samples from this particular patient, virtually all the insulin immunoreactivity was present in the form of this high molecular weight protein. In other sera significant amounts of insulin, but not proinsulin, were detected. However, further studies have suggested that the high molecular weight component may be an insulin–protein complex, which could be dissociated after acidification or incubation with urea[84]. Nunes-Correa *et al.*[85] studied a patient who suffered from severe hypoglycaemic attacks, presumably on the basis of an islet cell adenoma. In addition to insulin and proinsulin, they found a third peak in the patient's serum with a molecular weight of 24 000. The material in this peak reacted with insulin antibodies and had approximately 50–100% of the biological activity of insulin.

In our studies on insulinoma patients, variable amounts of material eluting in the void volume of Bio-Gel P-30 columns which react with insulin antibodies have been detected. The origin and significance of this component is still uncertain and further work is necessary in order to characterise it fully.

Circulating proinsulin — levels in normal subjects and disease states

The mean fasting proinsulin level in normal subjects is 0.16 ± 0.2 ng/ml[86]. In studies using a human insulin standard for measurement of proinsulin, PLC comprises approximately 15% of the total immunoreactive insulin concentration (range 0–22%). After oral glucose, the levels of proinsulin rise slowly and peak later than insulin. When expressed as a percentage of the insulin concentration, a decline from the fasting value is observed during the first 15 to 60 minutes. Thereafter proinsulin contributes an increasing amount to the immunoreactive insulin level. Obese patients have higher fasting concentrations of both proinsulin and insulin, so that the relative proportions of the two polypeptides are generally in the same range observed in healthy subjects[87].

The possibility that an altered proinsulin:insulin ratio might occur in patients with diabetes mellitus has been considered by a number of investigators. However, patients with mild diabetes characterised only by glucose intolerance, had normal proinsulin levels[87 88]. Duckworth *et al.*[89]

have reported that elderly, obese patients with carbohydrate intolerance have significantly elevated PLC concentrations. In contrast, children with mild diabetes and elevated immunoreactive insulin levels have proinsulin values within the normal range[90].

A number of other conditions are characterised by an elevated proinsulin:insulin ratio. These include patients with severe hypokalaemia of diverse aetiologies[91]. In severe diabetics and other patients with low fasting insulin levels, a similar situation may be found[92]. The absolute concentration and percentage PLC in the basal state in patients with chronic renal failure are markedly elevated and the values may overlap

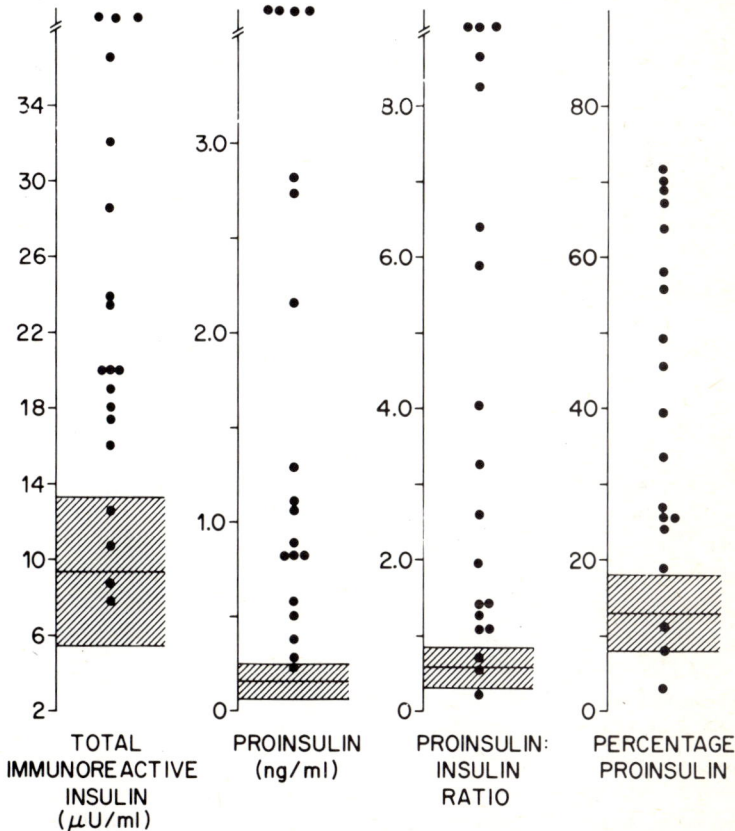

Figure 1.8 Basal total immunoreactive insulin (IRI), proinsulin-like components, proinsulin: insulin ratio and percentage proinsulin in 19 patients with islet cell tumours. The mean $\pm$ S.D. of 46 control subjects is shown in the hatched area. (Reproduced from Rubenstein, *et al.*, 1973.)

17

those found in subjects with islet cell tumours[86]. This situation arises because of the critical role of the kidney as the major organ involved in proinsulin degradation[93].

The major clinical significance of a raised serum proinsulin concentration has been in the diagnosis of pancreatic islet cell tumours[94 95]. We have studied 17 patients with β-cell adenomas and two with carcinomas[94]. The absolute basal PLC concentrations varied between 0.23 and 17.48 ng/ml. Only three patients overlapped the normal range (0.039–0.45 ng/ml). The percentage PLC ranged from 2.9 to 71 (normal values 4.6–22.8%) and only four of the insulinoma group fell within the range of the control subjects (Figure 1.8). These results are similar to those of Sherman *et al.*[95] who described three of 21 islet cell tumour patients with basal IRI concentrations within their normal range, but only one with a normal proinsulin concentration. Four subjects had percentage PLC which overlapped their controls. The findings in these two studies are representative of the conclusions in a number of other reports[88 96].

Metabolism of proinsulin

Proinsulin comprises 2–9% of the immunoreactive insulin-like material in normal pancreas[97 98]. This value is similar to that found in the portal vein of man[99], but much lower than in peripheral serum. This discrepancy can be explained by the slower metabolism of proinsulin compared to insulin. Sonksen *et al.*[100] infused porcine insulin and proinsulin into healthy subjects and showed a mean metabolic clearance rate of 13.3 ml/kg/min for insulin and 3.1 ml/kg/min for proinsulin with mean half-lives of 4.4 and 25.6 min respectively. The half disappearance time of endogenous proinsulin (18–25 min) was markedly slower than that of insulin (3–4 min) in three patients following removal of their islet cell tumours[101] (Figure 1.9).

In contrast to the difference in their MCR and half disappearance times, the renal disposition of proinsulin and insulin in rats is similar, being characterised by high extraction and very low urinary clearances[93]. The renal arteriovenous differences of proinsulin and insulin average 36 and 40% respectively, and are linearly related to their arterial concentrations between 2 and 25 ng/ml. The fractional urinary clearance never exceeded 0.6%, indicating that more that 99% of the amount filtered is sequestered in the kidney. On the other hand, removal of bovine proinsulin by the isolated perfused rat liver is considerably slower than that insulin, at both high and low concentrations[102 103].

A protease which is relatively specific for insulin, and does not significantly degrade proinsulin, has been described by Brush[104] who

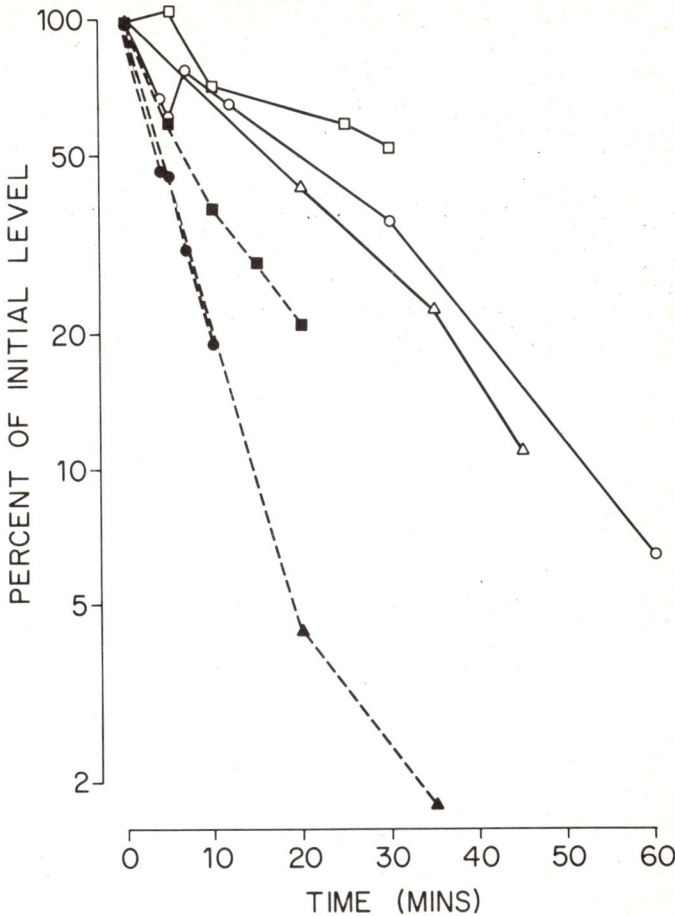

Figure 1.9 Disappearance of endogenous insulin (closed symbols) and proinsulin (open symbols) in three patients with islet cell tumors. The tumors were removed at 0 minutes and the subsequent fall in the peripheral serum insulin and proinsulin concentrations during the next 5 –60 minutes was followed (from Starr and Rubenstein, 1974).

isolated the enzyme from muscle, and by Burghen *et al.*[105] who found a similar enzyme in liver. The relative ability of different tissues to degrade insulin and proinsulin varies considerably. Only pancreas and kidney homogenates degrade immunoreactive·proinsulin at a rate greater than 10% that of insulin[106]. Whether or not this particular enzyme (ISP) is important in the degradation of insulin under physiological conditions is still uncertain. The problem of defining the specific cellular site where insulin degradation takes place under physiological circumstances is important and will undoubtedly require innovative experiments to solve.

Serum C-peptide

Because insulin and the C-peptide are derived from one molecule of proinsulin and are stored and released from the pancreas essentially in equimolar quantities, the serum concentrations of these peptides should correlate well. This supposition has been confirmed[107 108], and certain clinical situations in which the determination of serum C-peptide levels is particularly useful have been recognised. However, availability of the human C-peptide immunoassay has been hindered by several factors. Because of the poor cross-reactivity between human C-peptide and all the other C-peptides which have been studied[16], human C-peptide must be used for standard, radioactive label, and as antigen for the production of antibodies. Furthermore, human C-peptide has a relatively low molecular weight (3021 daltons) and thus tends to be a poor antigen. Even when coupled to large proteins such as bovine serum albumen, immunisation with C-peptide does not consistently result in a high-titre antiserum in either rabbits or guinea pigs. It is possible that a lack of rigid secondary or tertiary structure[13 108a] renders this peptide poorly immunogenic.

A human C-peptide radioimmunoassay was initially developed using C-peptide extracted from human pancreases removed at autopsy[109]. Recently, however, antisera have been produced more readily from synthetic C-peptide[110]. In order to iodinate the C-peptide to high specific activity, a tyrosyl residue must be added because there is no tyrosine or histidine residue in any of the known natural peptides. Tyrosyllated C-peptides have also been prepared synthetically by Yanaihara and his co-workers. Minor differences in immunoreactivity between endogenous and synthetic human C-peptide make it preferable to use the endogenous peptide as the standard. Because the entire C-peptide structure is present within the proinsulin molecule, proinsulin cross-reacts in the C-peptide assay. For this reason, the values have been referred to as C-peptide reactivity (CPR) rather than C-peptide. With present antisera, proinsulin displaces the tracer about 1/3 to 1/10 as well, on a weight basis, as C-peptide, and as C-peptide is present in approximately ten times higher concentration than proinsulin in fasting serum (2 ng/ml vs. 0.2 ng/ml), less than 3% of measured C-peptide immunoreactivity is attributable to proinsulin. Nevertheless, in sera containing higher amounts of proinsulin, the CPR level will not accurately reflect the C-peptide concentration.

Metabolism of C-peptide

The metabolic clearance rate (MCR) of bovine C-peptide is 4.6 ± 0.2 ml/min in the rat, which is less than that of either bovine insulin (16.4 ± 0.4 ml/min) or proinsulin (6.7 ± 0.3 ml/min)[93]. The MCR is independent of plasma concentrations over the range of 1–15 ng/ml. Renal removal of C-peptide from the circulation accounts for 69% of its MCR, while the renal contribution to insulin metabolism is only 33%. This difference is attributed to the liver having a significant role in the catabolism of insulin but not C-peptide. The kidney, on the other hand, is the major organ involved in the metabolism of the C-peptide.

C-peptide, together with insulin and proinsulin, is secreted into the portal circulation and passes through the liver before entering the peripheral blood. In order to relate peripheral concentrations to β-cell secretion, simultaneously determined portal and peripheral blood levels were compared in subjects whose portal blood was obtained by umbilical vein catheterisation prior to operation[99]. In general, both portal and peripheral serum concentrations of C-peptide were greater than those of insulin on a molar basis, but at times of peak secretion insulin and C-peptide were present in portal blood in nearly equimolar quantities. It is therefore likely that the slower MCR of C-peptide, together with the greater hepatic extraction of insulin, provides an adequate explanation for the higher peripheral molar concentrations of C-peptide compared to insulin.

Circulating C-peptide — levels in normal subjects and disease states

Because the insulin immunoassay does not distinguish endogenously secreted insulin from exogenously administered bovine or porcine insulin in insulin requiring diabetic patients, measurement of CPR has proven to be extremely helpful in assessing their β-cell secretory function. The assay for human C-peptide is not affected by small amounts of bovine or porcine proinsulin which may be present as impurities in commercial insulin preparations[11]. Furthermore, the development of circulating insulin antibodies in diabetic patients does not affect the C-peptide assay. Estimation of this peptide thus provides a measure of endogenous insulin secretion in these patients. Nevertheless, the results must be interpreted with caution because the circulating insulin antibodies may bind significant amounts of endogenously secreted proinsulin, which cross-react in the C-peptide assay[107 111].

Our understanding of the pathogenesis of diabetes mellitus has been enhanced by sequential studies of insulin levels in non-insulin requiring diabetic patients. The C-peptide immunoassay has permitted similar studies in insulin-dependent patients. Both insulin and C-peptide levels are very low or unmeasurable during episodes of ketoacidosis in adult onset diabetes. Following recovery from ketoacidosis, partial return of β-cell function occurs as assessed by increases in CPR[111]. These studies indicate that diabetic ketoacidosis does not necessarily imply that irreversible β-cell damage has occurred, and suggest that functional impairment as well as cellular destruction of β-cells may underlie this syndrome. Similar studies in patients with juvenille-onset diabetes mellitus[111] have confirmed the absence of β-cell secretion during periods of severe hyperglycaemia and ketonaemia (Figure 1.10). During the phase of clinical remission, the so-called 'honeymoon period' or 'Brush effect,' C-peptide secretion may resume to a variable extent. Eventual clinical relapse in these patients is again associated with a decline in serum CPR levels.

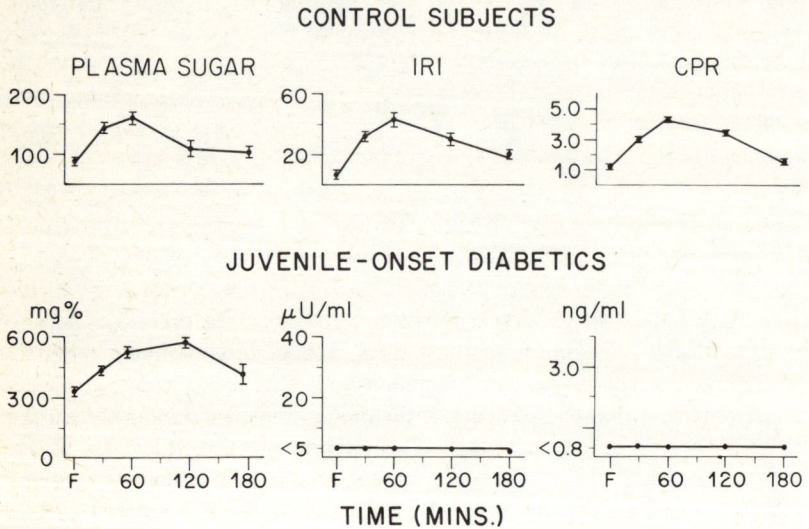

Figure 1.10 Plasma sugar, immunoreactive insulin (IRI) and C-peptide (CPR) levels during an oral glucose tolerance test in control subjects (above) and five juvenile-onset diabetics at the time of diagnosis of their disease (below).

The ability to measure serum C-peptide has also permitted the assessment of β-cell secretory capacity in insulin-requiring patients who have developed circulating insulin antibodies. As an important practical consideration, withdrawal of exogenous insulin is not necessary before evaluation of β-cell function is undertaken, because C-peptide levels specifically reflect endogenous insulin secretion. Moreover, the effect of insulin administration on β-cell secretory patterns can be determined, as well as the role of endogenous secretion in modifying the requirement for exogenous insulin.

The presence of residual β-cell secretory activity has been demonstrated in some adult-onset[107] and juvenile-onset[104] insulin-requiring diabetics. The CPR in these patients consists of both C-peptide and proinsulin. The proinsulin is largely bound to circulating insulin antibodies, accounting for the high proportion of proinsulin in the basal state. Following glucose, however, both C-peptide and proinsulin may rise in some patients, indicating an appropriate β-cell response to the glycaemic stimulus. In contrast, most ketosis-prone juvenile diabetics who have been treated with insulin for longer than five years have no measurable CPR in either the basal or post-stimulatory state. However, serum CPR is present in a number of juvenile-onset patients during the first years of their disease[112]. This finding suggests that the loss of β-cell secretory capacity in juvenile-onset diabetes is not abrupt, but continues for several years after the diabetes becomes clinically manifest.

Comparative studies of stable and unstable diabetics have been facilitated by C-peptide estimations. Adult insulin requiring patients, who were classified as stable on the basis of diurnal plasma and urine glucose variability, had greater CPR levels in both the basal state and following stimulation with either oral glucose or arginine[113] than unstable patients. In the latter group, basal CPR levels were at the lower limit of sensitivity of the assay, and no response to either glucose or arginine was detected.

The diagnosis of hypoglycaemic disorders has been greatly facilitated by measurements of serum insulin. However, difficulties arise in the evaluation of hypoglycaemia in diabetic patients who have been treated with insulin injections, for circulating insulin antibodies in these patients interfere with the insulin assay. In such patients, estimation of C-peptide may be used instead of insulin to document endogenous hyperinsulinism. Therefore, although insulinomas and nesidioblastosis are rare in diabetics, they can be distinguished from other causes of spontaneous, fasting hypoglycaemia such as mesenchymal tumours or hepatic or renal failure, as well as from non-insulin mediated disorders such as alcohol hypoglycaemia by this technique. Another difficult diagnostic problem in both diabetic and non-diabetic patients is distinguishing hyperinsulinism due to endogenous secretion from that due to the surreptitious injection of exogenous insulin. This distinction may be made by the simultaneous measurement of both serum IRI and CPR when the patient is hypoglycaemic. Because endogenously secreted insulin is accompanied by the liberation of C-peptide, while exogenous insulin will suppress endogenous secretion and hence lower the serum C-peptide concentration, serum IRI and CPR will show similar elevations in pancreatic hyperinsulinism, while CPR will be low in relationship to the IRI following injection of exogenous insulin. This observation has been of great help in diagnosing several cases where the patients were surreptitiously administering insulin to themselves[114].

Acknowledgements

Various aspects of this research was supported by grants from the Juvenile Diabetes Foundation, USPHS (AM-13914 and AM-13941) and the University of Chicago Diabetes–Endocrinology Centre (Am-17046). David Horwitz is a recipient of a Research and Development Award from the American Diabetes Association and also received support from the Louis Block Fund of the University of Chicago.

References

1. Steiner, D. F. and Oyer, P. E. (1967). The biosynthesis of insulin and a probable precursor of insulin by a human islet cell adenoma. *Proc. Natl. Acad. Sci US,* **57,** 473
2. Jacobson, M. R. and Baltimore, E. (1968). Morphogenesis of poliovirus. I. Association of the viral RNA with coat protein. *J. Mol. Biol., 33,* 369
3. Kiehn, E. D. and Holland, J. J. (1970). Synthesis and cleavage of enterovirus polypeptides in mammalian cells. *J. Virol.,* **5,** 358
4. Geller, D. M., Judah, J. D. and Nicholls, M. R. (1972). Interacellular distribution of serum albumin and its possible precursors in rat liver. *Biochem, J.,* **127,** 865
5. Judah, J. D., Gamble, M. and Steadman, J. H. (1973). Biosynthesis of serum albumin in rat liver. Evidence for the existence of "proalbumin." *Biochem. J.,* **134,** 1083
6. Bornstein, P. (1974). The biosynthesis of collagen. *Ann. Rev. Biol. Chem.,* **43,** 567
7. Steiner, D. F. (1967). Evidence for a precursor in the biosynthesis of insulin. *Trans. N. Y. Acad. Sci., Ser. II, 30,* No. 1, 60
8. Steiner, D. F., Kemmler, W., Clark, J. L. Oyer, P. E. and Rubenstein, A. H. (1972). The biosynthesis of insulin. In: *Handbook of Physiology,* Endocrinology I, p. 175 (D. F Steiner and N. Freinkel, eds.) (Baltimore: The Williams and Wilkins Co)
9. Steiner D. F., Peterson, J. D., Tager, H. S., Emdin, S. O., Ostberg, Y. and Falkmer, S. (1973). Comparative aspects of proinsulin and insulin structure and biosynthesis, In: *Proc. of Eighth Cong. of the Interntl. Diabetes Fed.,* Brussels. July,1973 (J. Pirart, and W. J. Malaissc, eds.) p. 119 Interntl. Cong. Ser. No. 312 (Amsterdam: Excerpta Medica)
10. Chance, R. E. Ellis, R. M. and Bromer, W. W. (1968). Porcine Proinsulin: characterization and amino acid sequence. *Science, 161,* 165
11. Steiner, D. F., Hallund, O., Rubenstein, A. H., Cho, S. and Bayliss, C. (1968). Isolation and properties of proinsulin, intermediate forms and other minor components from crystalline bovine insulin. *Diabetes, 17,* 725
12. Nolan, C., Margoliash, E., Peterson, J. D. and Steiner, D. F. (1971). The structure of bovine proinsulin, *J. Biol. Chem.,* **246,** 2780

13. Frank, B. H. and Veros, A. J. (1968). Physical studies on proinsulin: association behavior and conformation in solution. *Biochem. Biophys. Res. Comm.*, **32,** 155

14. Steiner, D. F., Clark, J. L., Nolan, C., Rubenstein, A. H., Margoliash, E., Aten, B. and Oyer, P. E. (1969). Proinsulin and the biosynthesis of insulin. *Rec. Prog. Hom. Res.*, **25,** 207

15. Rubenstein, A. H., Steiner, D. F., Cho, S., Lawrence, A. M. and Kirsteins, L. (1969). Immunological properties of bovine proinsulin and related fractions. *Diabetes,* **18,** 598

16. Rubenstein, A. H., Mako, M., Welbourne, W. P., Melani, F. and Steiner, D. F. (1970). Comparative immunology of bovine, porcine, and human proinsulin and C-peptides. *Diabetes,* **19,** 546

17. Gliemann, J. and Sorensen, H. H. (1970). Assay of insulin-like activity by the isolated fat cell method, IV. The biological activity of proinsulin. *Diabetologia,* **5,** 499

18. Narahara, H. T. (1972). Biological activity of proinsulin. In: *Insulin Action,* p. 63 (I, Fritz, ed.) (New York: Academic Press)

19. Lazarus, N. R., Panhos, J. E., Tanese, T., Michaels, L., Gutman, R. and Recant, L. (1970). Studies on the biological activity of porcine proinsulin. *J. Clin. Inv.,* **49,** 487

20. Frank, B. H. and Veros, A. J. (1970). Interaction of zinc with proinsulin. *Biochem. Biopys. Res. Comm.,* **38,** 284

20a. Steiner, D. F. (1973). Cocrystallization of proinsulin and insulin. *Nature,* **243,** 528

21. Low, B. W., Fullerton. W. W. and Rosen, L. S. (1974). Insulin/proinsulin, a new crystalline complex. *Nature,* **248,** 339

22. Moskalewski, S. (1965). Isolation and culture of the islets of Langerhans of the guinea pig. *Gen. Comp. Endocrin.,* **5,** 342

23. Lacey, P. E. and Kostianovsky, M. (1967). Method for the isolation of intact islets of Langerhans from the rat pancreas. *Diabetes,* **16,** 35

24. Steiner, D. F., Cunningham, D. D., Spigelman, L. and Aten, B. (1967). Insulin biosynthesis: evidence for a precursor. *Science,* **157,** 697

25. Tung, A. K. and Yip, C. C. (1968). The biosynthesis of insulin and 'proinsulin' in fetal bovine pancreas. *Diabetologia,* **4,** 68

26. Tanese, T., Lazarus, N. R., Devrim, S. and Recant, L. (1970). Synthesis and release of proinsulin and insulin by isolated rat islets of Langerhans. *J. Clin. Invest.,* **49,** 1394

27. Clark, J. L. and Steiner, D. F. (1969). Insulin biosynthesis in the rat: demonstration of two proinsulins. *Proc, Natl. Acad. Sci US,* **62,** 278

28. Markussen, J. (1971). Mouse insulins—separation and structures. *Int. J. Protein Research,* **3,** 149

29. Markussen, J. and Sundby, F. (1972). Rat proinsulin C-peptides. Amino acid sequences. *Eur. J. Biochem.,* **25,** 153

30. Sundby, F. and Markussen, J. (1972). Rat proinsulins and C-peptides isolation and amino-acid compositions. *Eur. J. Biochem.,* **25,** 147

31. Tager, H. S. and Steiner, D. F. (1972). Primary structures of the proinsulin connecting peptides of the rat and the horse. *J. Biol. Chem.,* **247,** 7936

32. Tager, H. S., Emdin, S. O., Clark, J. L. and Steiner, D. F. (1973). Studies on the conversion of proinsulin to insulin, II. Evidence for a chymotrypsin-like cleavage in the connecting peptide region of insulin precursors in the rat. *J. Biol. Chem.,* **248,** 7936

33. Grant, P. T. and Coombs, T. L. (1971). Proinsulin, a biosynthetic precursor of insulin In: *Essays in Biochemistry,* Vol. 6, p. 69 (P. N. Campbell and G. D. Greville, eds.) (London: Academic Press)

34. Trakatellis, A. C. and Schwartz, G. P. (1970). Biosynthesis of insulin in anglerfish islets. *Nature,* **225,** 548

35. Emdin, S., Peterson, J. D., Coulter, C. L., Ostberg, Y., Falkner, S. and Steiner, D. F. (1973). The structure and biosynthesis of insulin in a primitive vertebrate, the cyclostome, *Myxine glutinosa, Ninth Interntl. Cong of Biochem., Abstracts,* Stockholm, 452

36. Yamaji, K., Tada, D. and Trakatellis, A. C. (1972). On the biosynthesis of insulin in anglerfish islets. *J. Biol. Chem.,* **247,** 4080

37. Kemmler, W., Steiner, D. F. and Borg, J. (1973). Studies on the conversion of proinsulin to insulin. III. Studies *in vitro* with crude secretion granule fraction isolated from islet of Langerhans *J. Biol. Chem.,* **248,** 4544

38. Chance, R. E. (1971). Chemical, physical, biological and immunological studies on porcine proinsulin and related polypeptides. In: *Proc. of the Seventh Congress of the Interntl. Diabetes Fed.* p. 292. (R. R. Rodriguez and J. Vallance-Owen, eds.) Amsterdam: Excerpta Medica)

39. Hard, L. (1944). The origin and differentiation of the alpha and beta cells in the pancreatic islets of the rat. *Amer. J. Anat.,* **75,** 369

40. Munger, B. L. (1958). A light and electron microscopic study of cellular differentiation in the pancreatic islets of the mouse. *Amer. J. Anat.,* **103,** 275

41. Jamieson, J. D. and Palade, G. E. (1967). Intracellular transport of secretory proteins in pancreatic exocrine cell. I. Role of peripheral elements of Golgi complex. *J. Cell. Biol.,* **34,** 177

42. Sorensen, R. L., Steffes, M. W. and Lindall, A. W. (1970). Subcellular localization of proinsulin to insulin conversion in isolated rat islets *Endocrinology,* **86,** 88

43. Tijoe, T. O. and Kroon, A. M. (1973). Identification of insulin synthesizing polysomes in isolated islets of Langerhans from rat pancreas. *FEBS Letters,* **33,** 225

44. Hew, C. L. and Yip, C. C. (1974). The identification of proinsulin synthesizing polysomes in fetal bovine pancreas. *Can. J. Biochem.,* **52,** 959

45. Yip, C. C. and Liew, C. C. (1973). Role of methionine in the initiation of the biosynthesis of bovine proinuslin. *Can. J. Biochem.,* **52,** 783

46. Steiner, D. F., Clark, J. L., Nolan, C., Rubenstein, A. H., Margoliash, E., Melani, F. and Oyer, P. E. (1970). The biosynthesis of insulin and some speculations regarding the pathogenesis of human diabetes. In: *The Pathogenesis of Diabetes Mellitus,*p. 123 *Proc. Thirteenth Nobel Symposium.* (Stockholm: Almqvist and Wiksell)

47. Howell, S. L. (1972). Role of ATP in intracellular translocation of proinsulin and insulin in the rat pancreatic β cell. *Nature New Biology,* **235,** 85

48. Orci, L., Lambert, A. E., Kanazawa, Y., Amherdt, M., Rouiller, C. and Renold, A. E. (1971). Morphological and biochemical studies of β cells of fetal rat endocrine pancreas in organ culture. Evidence for proinsulin biosynthesis. *J. Cell Biol.,* **50,** 565

49. Howell, S. L., Kostianovsky, M and Lacy, P. E. (1969). Beta granule formation in isolated islets of Langerhans: a study by electron microscopic radioautography. *J. Cell Biol.,* **42,** 695

50. Steiner, D. F., Kemmler, W. Tager, H. S. and Peterson, J. D. (1974). Proteolytic processing in the biosynthesis of insulin and other proteins. *Fed. Proc.,* **33,** 2105

51. Kemmler, W., Peterson, J. D. and Steiner, D. F. (1971). Studies on the conversion of

proinsulin to insulin. I. Conversion *in vitro* with trypsin and carboxypeptidase B. *J. Biol. Chem.*, **246**, 6786

52. Kemmler, W. and Steiner, D. F. (1970). Conversion of proinsulin to insulin in a subcellular fraction from rat islets *Biochem. Biophys Res. Comm.*, **41**, 1223

52a. Zühlke, H., Jahr, H., Schmidt, S., Gottschling, D. and Wilke, B. (1974). Catabolism of proinsulin and insulin. Proteolytic activities in Langerhans islets of rat and mice pancreas *in vitro*. *Acto Biol. Med. Germ.*, **33**, 407

53. Sun, A. M., Lin, B. J. and Haist, R. E. (1973). Studies on the conversion of proinsulin to insulin in the isolated islets of Langerhans in the rat. *Canad. J. Physiol. Pharmacol.*, **51**, 175

54. Greider, M. H., Howell, S. L. and Lacy, P. E. (1969)). Isolation and properties of secretory granules from rat islets of Langerhans II. Ultrastructure of the beta granule. *J. Cell Biol.*, **41**, 162

55. Falkmer, S (1971). Sulfhydryl compounds and heavy metals in islet morphology and metabolism. In: *Proc. Seventh Congress of the Interntl. Diabetes Fed.* p. 219. (R. R. Rodriguez and J. Vallance-Owen, eds.) (Amsterdam: Excerpta Medica)

56. Logothetopoulos, J., Maneko, M., Wrenshall, G. A. and Best, C. H. (1964). Zinc, granulation, and extractable insulin of islet cells following hyperglycemia or prolonged treatment with insulin. In: *The Structure and Metabolism of the Pancreatic Islets*. Wenver-Gress Center Int Symp. Series, Volume 3, p. 333. (Oxford: Pergamon Press)

57. Smith, L. F. (1967). Amino acid sequences of insulins, *Diabetes*, 21 (Suppl. 2), 457

58. Peterson, J. D., Coulter, C. L., Steiner, D. F., Emdin, S. O. and Falkmer, S. (1974). Structural and crystallographic observations on hagfish insulin. *Nature*, **251**, 239

59. Grant, P. T., Coombs, T. L. and Frank, B. H. (1972). Differences in the nature of the interaction of insulin and proinsulin with zinc. *Biochem. J.*, **126**, 433

60. Lin, B. and Haist, R. E. (1975). Effects of some modifiers of insulin secretion on insulin biosynthesis. *Endocrinology*, **92**, 735

61. Hellman, B., Idahl, L., Lernmark, A. and Sehlin, J. (1974). The pancreatic β-cell recognition of insulin secretagogues. Comparison of glucose with glyceraldehyde isomers and dihydroxyacetone. *Arch. Biochem. Biophys.*, **162**, 448

62. Pipeleers, D. G., Marichal, M., and Malaisse, W. J. (1973). The stimulus-secretion coupling of glucose-induced insulin release. XV. Participation of cations in the recognition of glucose by the beta-cell. *Endocrinology*, **93**, 1012

63. Asplund, K. (1973). Effects of glucose upon insulin biosynthesis in fetal and newborn rats. *Horm. Metab. Res.*, **5**, 410

64. Asplund, K. and Hellerstrom, C. (1972). Glucose metabolism of pancreatic islets isolated from neonatal rats. *Horm. Metab. Res.*, **4**, 159

65. Creuzfeldt, W., Creuzfeldt, C., Frerichs, H., Perings, E., Sickinger, K. (1969). The morphological substrate of the inhibition of insulin secretion by diazoxide. *Horm. Metab. Res.*, **1**, 53

66. Stauffacher, E., Orci, L., Amherdt, M., Burr, I. M., Balant, L., Froesch, E. R., Renold, A. E. (1970). Metabolic state, pancreatic insulin content and β-cell morphology of normogycemic spiny mice (Acomys cahirinus): indications for an impairment of insulin secretion. *Diabetologia*, **6**, 330

67. Permutt, M. A. and Kipnis, D. M. (1972). Insulin biosynthesis II. Effect of glucose on ribonucleic acid synthesis in isolated rat islets. *J. Biol. Chem.*, **247**, 1200

68. Permutt, M. A. and Kipnis, D. M. (1972). Insulin biosynthesis. I. On the mechanism

of glucose stimulation. *J. Biol. Chem.,* **247,** 1194

68a. Andersson, A. (1975). Synthesis of DNA in isolated pancreatic islets maintained in tissue culture. *Endocrinology,* **96,** 1051

69. Steiner, D. F., Cho, S., Oyer, P. E., Terris, S., Peterson, F. D. and Rubenstein, A. H. (1971). Isolation and characterization of proinsulin C-peptide from bovine pancreas. *J. Biol. Chem.,* **246,** 1365

70. Rubenstein, A. H., Clark, J. L., Melani, F. and Steiner, D. F. (1969). Secretion of proinsulin C-peptide by pancreatic β cells and its circulation in blood. *Nature,* **224,** 697

71. Dayhoff, M. O. (ed.) (1973). *Atlas of Protein Sequence and Structure,* Vol. 5, Suppl. 1. Biomedical Research Foundation, Bethesda, Maryland

72. Oyer, P E., Cho. S., Peterson, J. D. and Steiner, D. F. (1971). Studies on human proinsulin. Isolation and amino acid sequence of the human pancreatic C-peptide. *J. Biol. Chem.,* **246,** 1375

73. Peterson, J. D., Nehrlich, S., Oyer, P. E. and Steiner, D. F. (1972). Determination of the amino acid sequence of the monkey, sheep and dog proinsulin C-peptides by a semi-micro edman degradation procedure *J. Biol. Chem.,* **247,** 4866

74. Dayhoff, M. O. (ed.) (1972). *Atlas of Protein Sequence and Structure,* Vol. 5, Biomedical Research Foundation, Bethesda, Maryland

75. Steiner, D. F. and Clark, J. L. (1968). The spontaneous reoxidation of reduced beef and rat proinsulins. *Proc. Natl. Acad. Sci. US,* **60,** 622

76. Roth, J., Gorden, P. and Pastan, I. (1968). "Big insulin"; new component of plasma insulin detected by immunoassay. *Proc. Natl. Acad. Sci. US,* **61,** 138

77. Rubenstein, A. H., Cho, S. and Steiner, D. F. (1968). Evidence for proinsulin in human urine and serum. *Lancet,* **i,** 697

78. Kitabchi, A. E., Duckworth, W. C., Brush, J. S. and Heinemenn, M. (1971). Direct measurement of proinsulin in human plasma by the use of an insulin-degrading enzyme. *J. Clin. Invest.,* **50,** 1792

79. Cresto, J. C., Lavine, R. L., Fink, G. and Recant, L. (1974). Plasma proinsulin: comparison of insulin specific protease and gel filtration assays. *Diabetes,* **23,** 505

80. Starr, J. I., Juhn, D. D., Rubenstein, A. H. and Kitabchi, A. E. (1975). Degradation of insulin in serum by insulin specific protease. *J. Lab. Clin. Med.* (In press)

81. Lazarus, N. R., Gutman, R. A., Panhos, J. C. and Recant, L. (1972). Biologically active circulating proinsulin-like materials from an islet cell carcinoma patient. *Diabetologia,* **8,** 131

82. Gorden, P., Freychet, P. and Nanken, H. (1971). A unique form of circulating insulin in human islet cell carcinoma. *J. Clin. Endocrinol.,* **33,** 983

83. Yalow, R. S. and Berson, S. A. (1973). "Big, Big, Insulin." *Metabolism,* **22,** 703

84. Sramkova, J., Pav, J. and Engelberth, O. (1975). Inordinately high levels of serum immunoreactive insulin in monoclonal immunoglobulinemia (On the problem of Big, Big, Insulin). *Diabetes,* **24,** 214

85 Nunes-Correa, J., Lowy, C. and Sonksen, P. H. (1974). Presumed insulinoma secreting a high molecular-weight insulin analogue. *Lancet,* **i,** 837

86. Mako, M. E., Block, M., Starr, J., Nielsen, D., Friedman, E. and Rubenstein, A. H. (1973). Proinsulin in chronic renal an hepatic failure: a reflection of the relative contribution of the liver and kidney to its metabolism. *Clin. Res.,* **21,** 631

87. Melani, F., Rubenstein, A. H. and Steiner, D. F. (1970). Human serum proinsulin. *J. Clin. Invest.,* **49,** 497

88. Gorden, P., Sherman, B. and Roth, J. (1971). Proinsulin-like components of circulating insulin in the basal state and in patients with islet cell tumors. *J. Clin. Invest.*, **50**, 2113

89. Duckworth, W. C., Kitabchi, A. E. and Heinemann, M. (1972). Direct measurement of plasma proinsulin in normal and diabetic subjects. *Amer. J. Med.*, **53**, 418

90. Rosenbloom, A., Starr, J. L., Juhn, D. and Rubenstein, A. H. (1975). Proinsulin in chemical diabetes of childhood and adolescence. *Diabetes,* (Submitted for publication)

91. Gorden, P., Sherman, B. M. and Simopoulos, A. P. (1972). Glucose intolerance with hypokalemia: an increased proportion of circulating proinsulin-like component. *J. Clin. Endocrinol.*, **34**, 235

92. Gorden, P., Hendricks, C. M. and Roth J. (1974). Circulating proinsulin-like component in man: increased proportion in hypoinsulinemic states. *Diabetologia*, **10**, 469

93. Katz, A. I. and Rubenstein, A. H. (1973). Metabolism of proinsulin, insulin and C-peptide in the rat. *J. Clin. Invest.*, **52**, 1113

94. Rubenstein, A. H., Mako, M. E., Starr, J. I., Juhn, D. J. and Horwitz, D. L. (1974). Circulating proinsulin in patients with islet cell tumors. *Proc. of Eighth Congress of the Interntl. Diabetes Fed.*, p. 736. (W. J. Malaisse, ed.) (Amsterdam: Ecerpta Medica)

95. Sherman, B. M., Pek, S., Fajans, S. S., Floyd, J. C. and Conn, J. W. (1972). Plasma proinsulin in patients with functioning pancreatic islet cell tumors. *J. Clin. Endocrinol.*, **35**, 271

96. Gutman, R. A., Lazarus, N. R., Penhos, J. C., Fajans, S. S. and Recant, L. (1971). Circulating proinsulin-like material in functioning insulinomas. *New Engl. J. Med.*, **284**, 1003

97. Rastogi, G. K., Letarte, J. and Fraser, T. R. (1970). Proinsulin content of pancreas in human fetusus of healthy mothers. *Lancet*, **i**, 7

98. Sando, H., Borg, J. and Steiner, D. F. (1972). Studies on the secretion of newly synthesized proinsulin and insulin from isolated rat islets of Langerhans. *J. Clin. Invest.*, **51**, 1476

99. Horwitz, D. L., Starr, J. I., Mako, M. E., Blackard, W. G. and Rubenstein, A. H. (1975). Proinsulin, insulin and C-peptide concentrations in human portal and peripheral blood. *J. Clin. Invest.*, **55**, 1278

100. Sonksen, P. H., Tomkins, C. V., Srivastava, M. C. and Nabarro, J. D. (1973). A comparative study on the metabolism of human insulin and porcine proinsulin in man. *Clin. Sci. Mol. Med.*, **45**, 633

101. Starr, J. I. and Rubenstein, A. H. (1974). Metabolism of endogenous proinsulin and insulin in man. *J. Clin. Endocrinol. Metab.*, **38**, 305

102. Rubenstein, A. H., Pottenger, L. A. M., Getz, G. S. and Steiner, D. F. (1972). The metabolism of proinsulin and insulin by the liver. *J. Clin. Invest.*, **52**, 512

103. Stoll, R. W., Touber, J. L., Menahan, L. H. and Williams, R. H. (1970). Clearance of porcine insulin, proinsulin and connecting peptide by the isolated rat liver. *Proc. Soc. Exptl. Biol. Med.*, **3**, 894

104. Brush, J. S. (1971). Purification and characterization of a protease with specificity for insulin from rat muscle. *Diabetes*, **20**, 151

105. Burghen, G. A., Kitabchi, A. E., and Brush, J. S. (1972). Characterization of a rat liver protease with specificity for insulin. *Endocrinology*, **91**, 633

106. Kitabchi, A. E. and Stentz, F. B. (1972). Degradation of insulin and proinsulin by various organ homogenates of rat. *Diabetes,* **21,** 1091

107. Block, D. B., Mako, D. E., Steiner, D. F. and Rubenstein, A. H. (1972). Circulating C-peptide immunoreactivity: studies in normals and diabetic patients. *Diabetes,* **21,** 1013

108. Horwitz, D. L., Starr, J. I., Rubenstein, A. H. and Steiner, D. F. (1973). Serum connecting peptide-an indicator of beta cell secretory function. *Diabetes,* **22,** Suppl. 1, 298

108a. Markussen, J. (1971). Structural changes involved in the folding of proinsulin. *Int. J. Protein Research,* **3,** 201

109. Melani, F., Rubenstein, A. H., Oyer, P. E. and Steiner, D. F. (1970). Identification of proinsulin and C-peptide in human serum by a specific immunoassay. *Proc. Natl. Acad. Sci. US,* **67,** 148

110. Yanaihara, N., Hashimoto, T., Yanaihara. C. and Sakura. N. (1970). Studies on the synthesis of proinsulin. I. Synthesis of partially protected tritriaconta-peptide related to the connecting peptide fragment of porcine proinsulin. *Chem. Pharm. Bull.* (Tokyo) **18,** 417

111. Block, M. B., Mako, M. E., Steiner, D. F. and Rubenstein, A. H. (1972). Diabetic Ketoacidosis: evidence for C-peptide and proinsulin secretion following recovery. *J. Clin. Endocrinol. Metab.,* **35,** 402

112. Block, M. B., Rosenfeld,. R., Mako, M. E., Steer, D. F. and Rubenstein, A. H. (1973). Sequential immunoreactivity. *New Engl. J. Med.,* **288,** 1148

113. Horwitz, D. L., Reynold, C., Molnar, G. D., Rubenstein, A. H. and Taylor, W. F. (1974). Stable and unstable diabetes–differences in endogenous insulin secretion as determined by C-peptide measurement.*Clin. Res.,* **22,** 471A

114. Couropmitree, C., Freinkel, N., Hagel, T. C., Horwitz, D. L., Metzger, B., Rubenstein, A. H. and Hahnel, K. (1975). Estimation of plasma C-peptide to diagnose factitious hyperinsulinism during "spontaneous" hypoglycemia in an insulin dependent diabetic. *Ann. Int. Med.,* **82,** 201

2
THE PANCREAS AND INSULIN RELEASE

S. J. H. Ashcroft and P. J. Randle

INTRODUCTION

The function of the β-cells of the islets of Langerhans is to synthesise insulin, to store it and, at the appropriate time, to release the required amount into the blood. β-cells contain the basic enzymic machinery for assembling the insulin molecule and for packaging it into its granular storage form, and receptors which permit the cells to monitor extracellular concentrations and to determine when, and how much, insulin should be released. They are also equipped with a system for coupling reception of the secretory signal with the process of secretion itself. In this chapter we review current knowledge and hypotheses concerned with these aspects of β-cell function. Consideration is given to the morphology of β-cells in relation to insulin synthesis and secretion; enumeration of those agents which modify the secretory process; the possible nature of stimulus–secretion coupling system(s) which regulate the secretory process; receptor system(s) for signal detection possessed by the β-cell. Finally we attempt to synthesise current views into a working model of insulin secretion and to explain its relevance to human diabetes.

MORPHOLOGY OF SECRETION

Anatomy of the β-cell[1][2]

In mammals the islets of Langerhans are scattered throughout the acinar pancreas. Despite large interspecies differences in total islet mass, the average size of mammalian islets of Langerhans remains roughly constant at around 1 μg dry weight per islet. The proportion of the insulin-producing β-cells in islets varies from species to species; mouse islets have a high content of β-cells (approx. 80%) and have therefore been widely used for biochemical studies. Individual β-cells may be in direct contact with each other at certain focal points on the plasma membrane (desmosomes). The islets are highly vascularised; the capillaries are separated from the β-cells by two basement membranes. Little is known of the role of these structures in the release process. Unmyelinated nerves are present in islets and synapses between nerve fibres and β-cells may occur.

A typical β-cell is surrounded by a plasma membrane and the following organelles are visible by microscopy : nucleus, mitochondria, ribosomes, lysosomes, endoplasmic reticulum, Golgi complex, secretory granules; microtubules and microfilaments have also been observed and the evidence for their involvement in secretion is discussed below. The characteristic secretory granules which contain virtually all of the insulin in β-cells have a central portion of variable shape and electron density, surrounded by a halo of low electron density and enclosed by a smooth membranous sac. The Golgi complex is usually near the cell nucleus and consists of parallel cisternae limited by smooth membranes; associated with the complex are varying numbers of vesicles[3]. In the area between the rough endoplasmic reticulum and the Golgi complex numerous microvesicles occur, particularly in cells actively engaged in insulin synthesis. Some of these appear to bud off from the rough endoplasmic reticulum: these elements have been suggested to be involved in the protected transfer of newly-synthesised insulin from the RER to the Golgi complex[4].

Insulin biosynthesis and storage

Insulin is synthesised as the single-chain precursor, proinsulin[5], in which a connecting peptide segment (CP) of about 30 amino acid residues links the C-terminal residue of the B-chain of insulin with the N-terminal residue of the A-chain[6]. Electron microscopy-autoradiography has shown that, following biosynthesis, proinsulin is transferred to the cisternae of the endoplasmic reticulum and then transported to the forming face of the Golgi complex[7][8]. Within the Golgi cisternae, proinsulin is concentrated into granule form, and at some stage during secretory granule formation proinsulin is converted into insulin by removal of the CP. The biosynthesis of proinsulin and its transfer to the Golgi region are energy–requiring processes: proteolysis to insulin however proceeds independently of energy metabolism. Current evidence suggests that conversion begins in the Golgi complex and continues in the maturing secretory granule prior to secretion. The CP and the basic amino acid residues are retained within the secretory granule together with the insulin. In most species the secretory granules also contain zinc.

Properties of isolated granules

Methods have been devised for the isolation of preparations of granules from islets[9] [10]. β-granules from microdissected pancreatic islets of obese, hyperglycaemic mice[9] showed maximum stability at pH 6 in media of low ionic strength. Granule stability was not affected by sucrose (50–320 mM). Sodium deoxycholate liberated all bound insulin; ATP and citrate also decreased granule stability. The stability of granules from islets of normal rats or mice was not affected by ATP. Granules from normal rat islets[11] were stable in 0.3 M sucrose and isotonic salt solutions; more stable in potassium rich media than in sodium rich media; granule stability was not affected by Ca^{2+}, Mg^{2+} or phosphate. Stability was greatest in the pH range 5–7 with an optimum pH of 6. No direct effects on granules of agents that affect insulin release in β-cells have been detected.

The mechanism of insulin release

The major if not the only route for the secretion of insulin is believed to be the process of exocytosis. In this process granules move to the periphery of the cell and into contact with the plasma membrane. At the point of contact a hiatus is produced through which the granule contents can be released into extracellular fluid. Such a mechanism was first postulated for insulin release by Lacy[12] on the basis of electron microscope studies of β-cells although there is evidence for this process in many other secretory cells, e.g. acinar pancreas, pituitary, adrenal medulla and mast cells. Although the electron microscope has provided many convincing images of β-cell granules apparently undergoing exocytosis, it is not easy to obtain quantitative data showing an increased rate of this process in cells actively secreting, nor to establish that this is the sole means of release. In contrast to the acinar pancreas, the β-cell secretes only a small proportion (<5%) of its stored insulin on *in vitro* stimulation or under physiological conditions. Because of this Findlay *et al.*[13] were not able to detect any increase in granule–membrane contacts in electron micrographs of β-cells during stimulation of insulin release with glucose or tolbutamide. The formation of cytoplasmic projections (microvilli) apparently resulting from fusion of granules and plasma membrane at neighbouring loci has been found by Lacy[14]; with the scanning electron microscope an increase in the number of microvilli in cells presumed to be β-cells in guinea-pig islets stimulated to secrete insulin was observed.

Further quantitative evidence for an increase in exocytosis in stimulated β-cells has been given by Orci *et al.*[4]; using the freeze-etching technique, a striking increase was detected in the number of stomata (presumed to correspond to secretory events) on the surface of β-cells from islets incubated under stimulatory conditions as compared with islets in the basal state. Orci *et al.*[15] have seen material similar to insulin granule contents in the extracellular space of stimulated foetal rat islets. Direct chemical evidence for exocytosis is provided by the finding that the connecting peptide, which is present in granules with insulin, is released from β-cells in the same 1:1 ratio to insulin that occurs in the granules[8]. Studies *in vitro* with islets incubated with radioactive amino acids have indicated that preferential release of newly-synthesised insulin does not occur; thus apparently all the insulin must pass through the granule stage before being secreted. Nevertheless, some studies have suggested that insulin release can occur from degranulated β-cells: these observations have been reviewed by Creutzfeldt[16] who finds them inconclusive evidence for insulin release by a non-exocytotic route.

Since during exocytosis the granule membrane is apparently incorporated into the plasma membrane, a mechanism must exist for a compensatory movement of membrane from the cell membrane to the cytoplasm in order that the cell surface does not increase indefinitely. Orci[17] has obtained evidence for an increase in the endocytotic uptake of exogenous peroxidase in stimulated β-cells. It may be significant that lysosomes and autophagic bodies are seen in β-cells: the latter often contain lamellated or whorled structures that may represent phospholipid residues from the breakdown of membranous organelles.

A major problem which has not yet been definitely resolved for any secretory cell type concerns the factors involved in the translocation of granules to the cell membrane during exocytosis. Matthews[18] has calculated that Brownian motion of granules in the cell sap may be a sufficiently rapid process to account for the speed of onset of secretion of insulin in response to a stimulus.

However such a calculation in no way rules out a more directed granule translocation system and persuasive evidence has accumulated for the participation of such an intracellular system, the morphological components of which are microtubules and microfilaments. Evidence for the participation of microtubules and microfilaments in secretion has come from electron microscopy of β-cells and from studies of the effects on insulin release of drugs which are known to affect the function and morphology of microtubules and microfilaments in other systems. Microtubules are tubular structures of 240 Å diameter, often of considerable length and found in such diverse sources as the mitotic spindle, cilia and flagella, and brain[19][20]. They are composed of aggregates of identical subunits of the protein tubulin. Tubulin is a dimer composed of two non–identical subunits each of molecular weight 55 000. It contains one molecule of tightly-bound guanosine di-or triphosphate per molecule of tubulin and also binds one molecule of the antimitotic agents colchicine and vinblastine at separate binding sites. The microtubule is a highly dynamic structure capable of rapid assembly and disassembly in response to various drugs and conditions. The binding of colchicine or vinblastine promotes depolymerisation of the microtubule, whereas the microtubules are stabilised by D_2O. Vinblastine differs from colchicine in that it also induces the formation of tubulin paracrystals. An important role for Ca^{2+} *in vivo* assembly and disassembly is suggested by reports of successful reassembly of microtubules from tubulin in the presence of a calcium chelating agent[21]. Tubulin can be phosphorylated *in vitro* by cyclic AMP-dependent protein kinase[22] but there is no proof that phosphorylation is involved in *in vivo* assembly or disassembly. In the β-cell, microtubules can be seen in electron micrographs[23]: the evidence for their involvement in insulin release is based on studies of the effects of the agents discussed above which interfere with microtubules[23-25]. Thus treatment of islets with colchicine or vinca alkaloids impairs insulin secretory responses. In β-cells from islets treated with vinca alkaloids dense masses can be seen in the electron microscope presumed to be the paracrystalline tubulin aggregates mentioned above. With both these agents insulin release is impaired rather than abolished. A marked time dependency of the effect of vincristine is seen[25]. With short exposure some enhancement of secretion is found which gives way to inhibition with more prolonged exposure. Associated with these changes in release is a gradual disappearance of microtubules and appearance of crystalline deposits on electron microscopy. However more dramatic effects on insulin release are seen in islets incubated in medium in which H_2O is progressively replaced with D_2O. At 50% D_2O the insulin secretory response to glucose is almost completely blocked.

Interpretation of these observations is clearly dependent on the specificity of the agents used. The stimulatory effects of glucose on insulin biosynthesis and [45]Ca uptake into the β-cell are not blocked by these agents[23]. There are no reports of gross metabolic derangements of β-cells induced by these agents. However actions of colchicine and vinblastine in other systems show that they may not be completely specific for microtubules. Thus colchicine inhibits nucleoside transport across the plasma membrane of various mammalian cell lines[26]. Vinblastine has been shown to precipitate proteins other than tubulin including muscle actin[27]. D_2O has been shown to inhibit excitation–contraction coupling in barnacle muscle fibres by interfering with release of Ca^{2+}[28]. Some caution must therefore be exercised in accepting a role for microtubules in insulin release.

A second component postulated to be involved in exocytosis is microfilaments. Contraction in a variety of cells including neurones and glia is believed to involve contractile microfilaments which can be disrupted by the fungal metabolite cytochalasin B[29]. The presence in the β-cell of a network of microfilaments 40–70 Å in diameter located just beneath the plasma membrane has been reported[30][31]. This filamentatous network, or 'cell web', was altered in glucose-stimulated β-cells exposed to cytochalasin B. Instead of a discrete band, the cell web appeared as large masses of closely packed microfilaments extending far into the cytoplasm. In parallel with the morphological changes, insulin release was enhanced. These findings are interpreted by Malaisse *et al.*[31] as suggesting that the cell web may restrict access of secretory granules to the plasma membrane. Again the specificity of the agent used is important. Cytochalasin B inhibits the ATPase activity of actomeromyosin[32], and interferes with glucose utilisation in a number of tissues[33], including pancreatic islets[34]. Moreover both inhibiting and stimulatory effects of cytochalasin B on insulin release can be demonstrated depending on the dose used, suggesting more than one site of action[35].

Possible ways in which microfilaments and microtubules could direct granule movement are numerous and, at the present time, all speculative. The microtubular system could be a passive network directing granules to the desired loci on the plasma membrane. In association with contractile microfilaments, the microtubules could be directly involved in the movement of granules. Orci *et al.*[30] suggest that the cell web might, in addition to a restrictive role, impart a final active impulse to secretory granules approaching the cell membrane.

Little attention has been given to the possible involvement of actomyosin-like contractile proteins in insulin secretion. Such proteins have been extracted from various secretory cells including platelets, cells of adrenal medulla and cells of mammalian brain. The protein isolated from mammalian brain has been called neurostenin, and it can be dissociated into an actin-like (neurin) and a myosin-like (stenin) component; the occurrence of this protein in the synaptosome fraction of brain has led to the suggestion of a key role for neurostenin in exocytotic discharge of transmitter substance at synaptic junctions: Berl *et al.*[36] hypothesise that at the site of contact of vesicles and presynaptic membrane, the combination of neurin and stenin results in a change in the membranes leading to opening of the vesicle and release of transmitter. It is noteworthy that colchicine, vinblastine and cytochalasin B all inhibit the Mg^{2+} stimulated ATPase activity of synaptosomal neurostenin[36]. Such a model could be applicable to β-cell exocytosis but there is no evidence for it. In fact nothing is known about the factors involved in the fusion of granule and plasma membranes and subsequent release of granule content. Matthews[18] has suggested that deletion of like charges on granules and plasma membranes may facilitate contact of granule and plasma membrane.

EFFECTORS OF INSULIN RELEASE

In considering mechanisms involved in the control of insulin release we have distinguished two broad classes of stimulatory agents. Firstly there are those agents that are themselves capable of eliciting insulin release: we shall refer to these as *initiators*. Secondly there are agents which themselves have little or no effect on insulin release in the absence of other agents, but which increase the secretory response to an initiator: we shall refer to such agents as *potentiators*.

Those agents which can elicit insulin release are as follows (for reviews see Randle and Hales[37], Malaisse[38], Grodsky[39]). Among physiological effectors, only glucose, mannose and leucine are undisputed initiators of insulin secretion. Pharmacological initiators include the sulphonylureas, *p*-chloromercuribenzoate, Ba^{2+} and an elevated K^+ concentration. Potentiators include other amino acids, agents which elevate cyclic $3',5'$-AMP concentration, (e.g. glucagon and caffeine) and sugars and sugar derivatives apart from glucose and mannose (see below). Insulin release is inhibited by a number of agents; the inhibition by mannoheptulose is specific for initiation by glucose and mannose and for potentiation based on initiation by these sugars. Absence of external Ca^{2+} blocks secretory responses to all agents except for Ba^{2+} and possibly, *p*-chloromercuribenzoate; adrenaline and diazoxide are effective inhibitors against a range of stimuli. Adequate β-cell ATP is also required for insulin secretory responses. Consequently agents which interfere with ATP synthesis reduce or abolish secretory responses.

Time course of insulin release

The kinetics of insulin release by the perfused rat pancreas have been extensively studied by Grodsky and his colleagues (for review see Grodsky[40]). The time course of the secretory response to glucose has been shown to be biphasic. An increased glucose concentration elicits within 0.5 − 1 min a rapid rise in the rate of insulin release which returns almost to basal within 5–10 min; this initial spike is followed by a more gradual increasing rise in the rate of insulin release. With tolbutamide as stimulus, the initial rapid release is followed by a constant but increased rate of release. Approximately 2–3% of the pancreatic insulin is released during the first phase of glucose stimulation and about 20% during the second phase on perfusion for 60 min. These observations have suggested that insulin may be stored in two compartments, the smaller of these being particularly labile to stimulants. The initial phase of release is ascribed to emptying of this labile compartment and the second phase to slower release of insulin from the large storage compartments. An additional effect of glucose is postulated to account for the increasing second phase of glucose-stimulated release: this additional effect may involve proinsulin and insulin biosynthesis or the release of insulin from the large compartment. Computer simulation has shown that a two-compartment model can reproduce the features of variously applied glucose and tolbutamide stimuli.

Kinetic studies have also suggested that the labile compartment is non-homogeneous and contains 'packets' of insulin with a Gaussian distribution of sensitivities to glucose[40 41]. This conclusion is derived from study of the amounts of first-phase insulin released in response to submaximal concentrations of glucose presented either as single continuous steps at various concentrations, or as staircase stimulations of increasing glucose concentrations. Essentially a homogeneous labile compartment would respond to different glucose concentrations with different maximal secretion rates: the *total* insulin released would however be the same. On the other hand a labile compartment with a range of threshold sensitivities would release varying total amounts of insulin depending on the number of packets whose threshold sensitivity was reached or exceeded. The data obtained support the latter model and for a full description of this hypothesis the reader is referred to Grodsky[40 41].

Whether there is any morphological basis for two insulin storage compartments is not known. The large compartment could represent typical storage granules, whilst the labile compartment might comprise β-granules located at sites especially favourable for release, e.g. aligned along microtubules. However, in the absence of direct evidence for compartmentation at the level of insulin storage, other explanations for biphasic insulin release cannot be excluded.

STIMULUS-SECRETION COUPLING

Involvement of Ca^{2+}

In common with other secretory cell types it is currently envisaged that the insulin release mechanism is activated by a rise in the cytosolic concentration of Ca^{2+}. Various lines of evidence support this key role for Ca^{2+}. The requirement for the presence of extracellular Ca^{2+} in order for stimulation of release to occur has been noted above. Curry *et al.*[42] found that the amount of insulin released by the perfused rat pancreas during a two minute glucose pulse was proportional to the medium Ca^{2+} concentration over the range 0.25 − 2 mM and reached a plateau at about 5 mM Ca^{2+}. Replacement of extracellular Ca^{2+} by Sr^{2+} permitted a normal stimulation of insulin release by glucose[43]. Replacement of Ca^{2+} by Ba^{2+} stimulated insulin release[44]. Beryllium could not replace Ca^{2+} and, when added in the presence of Ca^{2+}, 2.5 mM Be^{2+} inhibited release, as did elevation of Mg^{2+} to 12 mM[43,44]. Ni^{2+} which inhibits contraction of cardiac muscle by interfering with the uptake and/or the action of Ca^{2+} on excitation–contraction coupling, is a potent inhibitor of insulin secretion[45]. Other agents which may be expected to interfere with Ca^{2+} uptake into the β-cell also inhibit insulin release. Thus local anaesthetics such as tetracaine block insulin release evoked by a number of stimuli, and in this laboratory Mrs S. Manley has found that ruthenium red which may inhibit Ca^{2+} movement across plasma and mitochondrial membranes is a powerful inhibitor of insulin release.

Electrophysiological measurements of β-cell membrane electrical activity have been carried out by Dean and Matthews[46–48]. Glucose induced action potentials which occurred in bursts separated by intervals of about 3–4 s. Individual action potentials lasted 25–30 msec with an interval of about the same duration. The mean membrane potential was also dependent on the glucose concentration, changing from −33 mV to −15 mV as the glucose concentration was increased from zero to 27.5 mM. Exposure of islets to calcium-free medium for 60 min abolished the ability of glucose to elicit action potentials, whereas an increase of Ca^{2+} to 7.7 mM increased the amplitude of the glucose-induced action potentials. Strontium could be substituted for Ca^{2+}, but manganese blocked the potentials and depolarised the cell. The action potentials were insensitive to changes in chloride concentration but their amplitude was increased when the Na$^+$ concentration was lowered to 25 mM. From their studies Dean and Matthews conclude that glucose-induced electrical activity in the β-cells is due primarily to the entry of Ca^{2+}.

Further direct evidence for a key role for Ca^{2+} has been provided by Malaisse in studies of the handling of $^{45}Ca^{2+}$ by isolated rat islets (for review see Malaisse[23]). The initial observation was that the retention of radioactivity in islets incubated with ^{45}Ca for 90 min and subsequently washed extensively was increased by glucose. In later experiments, islets were preloaded with ^{45}Ca and then perfused under various conditions with continuous monitoring of the rate of efflux of ^{45}Ca. On stimulation with glucose an initial fall in the rate of efflux was followed by a marked increase in ^{45}Ca efflux. This secondary rise was not seen if the glucose stimulation was applied under conditions under which insulin release was blocked (e.g. absence of extracellular Ca^{2+} or presence of D_2O). The secondary rise was therefore attributed to ^{45}Ca associated with the secretory granules and released concomitantly with insulin. Glucose was postulated to increase intracellular Ca^{2+} by inhibiting its outward transport across the plasma membrane. A direct relationship was found between the rate of insulin release and the accumulation of ^{45}Ca when the former was varied by various stimulatory and inhibitory agents. An important exception to this relationship was that agents that increase the concentration of cAMP in islets, such as theophylline, enhanced glucose-stimulated insulin release (see below) but did not modify glucose-stimulated ^{45}Ca accumulation. However perfusion experiments showed that theophylline increased the rate of ^{45}Ca efflux from islets preloaded with ^{45}Ca and that this effect was much greater in the absence than in the presence of glucose. Malaisse[24] therefore suggests that the effect of theophylline is to induce a movement of Ca^{2+} from some intracellular organelle to the cytosol. In the absence of glucose, most of this Ca^{2+} is lost from the cell, thus accounting for the failure of theophylline to sustain insulin release in the absence of glucose (glucose prevents the efflux of Ca^{2+} from the cell).

Involvement of cAMP

There is persuasive evidence that the release of insulin may be modified by changes in the β-cell concentration of cAMP (for review see Randle and Hales[37], Malaisse[38]). Under appropriate conditions, agents which stimulate adenylcyclase (glucagon or β–adrenergic effectors) or inhibit cAMP phosphodiesterase (caffeine or theophylline) increase rates of insulin release. cAMP itself, or its dibutyryl analogue, has also been reported to stimulate insulin. However the specificity of these effects is doubtful since AMP, ADP and ATP have also been reported to have stimulatory effects. It has recently become possible to study more directly the control of cAMP concentration in pancreatic islets under various conditions, and the

The pancreas and insulin release

Sensitive assays for cAMP have permitted measurement of its concentration in pancreatic islets under various conditions, and the properties of adenylcyclase, phosphodiesterase and cAMP-dependent protein kinase have been examined. A major goal of these investigations has been to establish whether cAMP may mediate the stimulatory effect of glucose on insulin release. Most of the evidence suggests that an elevation of β-cell cAMP concentration is neither a necessary nor a sufficient condition for stimulation of insulin release, but that such a rise may amplify the secretory response to an initiator of release. Thus no effects of a stimulatory glucose concentration could be detected on cAMP concentration in rat or mouse islets[49][50] although insulin release was increased. Glucose was also without detectable effect on the activity in islets of phosphodiesterase[51], adenylcylase[52] or cAMP-dependent protein kinase[53]. On the other hand caffeine and isobutylmethylxanthine which markedly elevates islet cAMP by inhibiting cAMP phosphodiesterase did not initiate insulin release, but potentiated effects of glucose[49-52]. Some increase in the islet concentration of cAMP with glucose has been detected by Charles et al.[54] and by Grill and Cerasi[55], but again the changes of cAMP did not correlate with rates of insulin release.

A possible role of cAMP in mediating sulphonylurea-stimulated insulin release has received attention. It has been shown that glibenclamide inhibits β-cell phosphodiesterase in extracts of mouse islets[51], activates adenylcylase[56] and increases cAMP-dependent protein kinase activity[53]. Hellman et al.[57] have shown that sulphonylureas are restricted to the extracellular space of pancreatic islets and presumably interact with some component of the plasma membrane: adenylcylase is situated in the membrane but the location of phosphodiesterase in β-cells is not known.

Islet adenylcyclase has been shown to be activated by fluoride, prostaglandins, glucagon, ACTH, secretin, pancreozymin and GTP and to be inhibited by adrenaline and diazoxide[52][56][58]. The presence of two phosphodiesterases with differing K_m values for cAMP in extracts of islets has been reported[51][59]. The low K_m activity is inhibited by methylxanthines and sulphonylureas and activated by imidazole[51]. cAMP-dependent protein kinase activity is increased in rat islets incubated with glucagon, methylxanthines, sulphonylureas, and decreased by adrenaline or diazoxide[53]; as in other tissues the enzyme is dissociable into catalytic and regulatory subunits[60].

These data suggest that the role of cAMP in the control of insulin release is that of a modulator whose concentration in the β-cell may determine the magnitude of the secretory response to initiators such as glucose.

The site of action of Ca^{2+} and cAMP on insulin release

Any suggestions as to the possible sites of action of Ca^{2+} and cAMP in insulin release are highly speculative at the present time. Rasmussen[61] draws attention to the widespread occurrence of Ca^{2+} and cAMP as coupling factors between cell excitation and response and discusses the relationship between the two. A model of the events in insulin secretion is proposed in which cAMP activates a protein kinase which phosphorylates a component of the microtubular–microfilamentous system and thereby confers on this system the ability to respond to Ca^{2+} : activation of this system by Ca^{2+} in some way then leads, perhaps by a contractile event, to movement of the secretory granule to the cell membrane, where an additional role of Ca^{2+} may be to facilitate the membrane–membrane interaction required for exocytosis to occur. Equally plausible however is to assign to Ca^{2+} the role of activating the secretion complex, either directly or perhaps via a phosphorylation reaction catalysed by a Ca^{2+}-dependent protein kinase, and the site of action of cAMP to intracellular organelles able to take up and release Ca^{2+}. It may well be that further progress in understanding these events will come from study of other secretory cells where larger amounts of available tissue will facilitate the identification and characterisation of subcellular components of the release process.

SIGNAL RECOGNITION AND TRANSDUCTION

The β-cell responds to two classes of physiological signals, namely substrates and hormones. The substrates are sugars, amino acids and lipid metabolites. Particular emphasis has been given to the effects of glucose on the β-cell in view of its importance in glucose homeostasis.

Glucose-induced insulin release

The experimental findings that any model of the β-cell glucoreceptor must account for are as follows. The receptor concerned with initiation shows considerable specificity. Of a large number of sugars and derivatives tested only glucose, mannose, D-glyceraldehyde and, to a small extent, glucosamine are able to initiate insulin release. The following sugars or sugar derivatives may act as potentiators. N-acetylglucosamine has been shown to potentiate insulin release from mouse islets with glucose, D-glyceraldehyde or leucine as initiators. D-fructose and D-galactose (rat only) potentiate with glucose as initiator. D-mannoheptulose potentiated release with D-glyceraldehyde as initiator[62][63]. The response to glucose is sigmoid and especially sensitive to glucose concentrations just above the normal fasting blood levels. These properties may be attributes of the glucoreceptor which in combination with glucose leads to activation of the coupling system to initiate insulin secretion.

Two models for the glucoreceptor have been proposed[64]. In the *regulator-site* model glucose acts as an allosteric modifier presumably of a cell surface receptor; combination with glucose may now be regarded as leading via changes in membrane permeability to an influx of Ca^{2+} which triggers exocytosis. In the *substrate-site* model glucose is metabolised by the β-cell resulting in a change in concentration of an intracellular metabolite or cofactor thus initiating influx of Ca^{2+} and exocytosis. In this model the glucoreceptor is an enzyme system controlling the rate of entry of glucose into metabolism in the β-cell. These models have provoked considerable controversy: the current position may be as follows.

Regulator-site model

The main arguments adduced in favour of the regulator-site model have been based on attempts to dissociate the metabolism of sugars and their ability to initiate insulin release. The substrate-site hypothesis requires that the speed of the secretory response and metabolism of the sugar should be similar. Matschinsky[65] has argued against the substrate-site model on the grounds that the change in concentration of glycolytic intermediates in rat islets in response to hyperglycaemia is too slow to account for the rapidity of the secretory response. This argument has not been supported in other studies. Panten[66] has shown extremely rapid changes in pyridine nucleotide fluorescence on exposure of islets to high glucose. Ashcroft *et al.*[67] demonstrated increases in hexose phosphate concentration in both mouse and rat islets within 5 min of increasing the extracellular glucose concentration. Matschinsky[34] has used iodoacetate as an inhibitor of glycolysis in an attempt to dissociate glycolysis from glucose-stimulated release. With 0.2 mM iodoacetate and with pyruvate as a respiratory substrate, glucose stimulated insulin release whilst lactate output (measured as an index of glycolysis) was very low. However the interpretation of these experiments is difficult to assess, since iodoacetate itself is able to stimulate insulin release in the presence of high glucose[68]. Moreover lactate output does not necessarily estimate glycolytic flux. An increase in glycolytic rate (measured as lactate output) in the presence of galactose or 3-0-methyl-glucose has been claimed by Matschinsky[35] to occur without a parallel change in insulin release; such effects of galactose and 3-0-methyl-glucose on islet glucose utilisation on oxidation however were not observed by Hellman *et al.*[69] nor by ourselves (unpublished observation). Yet another line of reasoning advanced in support of the regulator-site has been the claim that insulin release was stimulated by the non-metabolised sugar galactose[70]. (Glucosamine and *N*-acetylglucosamine which have sometimes been referred to as non-metabolised sugars[70] are metabolised by mouse and rat islets[63]. Galactose however is not appreciably oxidised by islets of either species[63]. Hence the stimulatory effect of galactose on insulin release claimed by Matschinsky[70] is of particular interest.) The finding has not been confirmed in a number of studies[62 71-74]. Direct evidence for the regulator-site model would be the detection of membrane-associated glucose-binding activity in extracts of islets coupled with evidence that the activity was not a glucose transport system. A simple calculation illustrates the potential difficulty of such a demonstration. If we imagine 20 000 glucoreceptor molecules per cell (fat cell insulin receptor density) and suppose that there are 1000 β-cells per islet, then 1000 islets (the maximum yield of current isolation procedures) could bind 2×10^{10} glucose molecules. Since the K_m glucose of the glucoreceptor is *c.* 5 mM, maximal glucose-binding would require 50 mM glucose. A solution of 50 mM glucose whose concentration would be changed appreciably (say by 10%) by binding of 2×10^{10} molecules would have a volume of $\sim 10^{-7}$ μl. One approach to this problem has been reported by Price[75] employing the technique of difference spectrometry. A perturbation of absorbance at 283 nm by islet membranes was observed on adding glucose or mannose. Effects of other sugars were not reported however and the possibility of this presumed sugar-protein interaction being attributable to sugar binding by a transport system is not excluded.

Substrate-site model

The substrate-site model attributes the insulin secretory response to an increased flux of glucose through metabolic pathways. As originally conceived, the model was concerned with the possibility that a metabolite of glucose may initiate the secretory process. An alternative possibility is that the flux of glucose into the cell, occasioned by changes in extracellular concentration and sustained by intracellular metabolism may facilitate ion fluxes and initiate the secretory process. These considerations led to the prediction that changes in the rate of glucose (or mannose) utilisation would be correlated with changes in the rate of secretion. This expectation has been realised in experiments in which glucose utilisation and insulin release were altered by varying extracellular glucose concentration[62]; in inhibition by mannoheptulose and glucosamine (which inhibit glucose phosphorylation)[62]; and in inhibition by iodoacetate which blocks glycolysis (S. J. H. Ashcroft, unpublished work). These studies have involved the development of methods for microassay of oxygen consumption[75]; glycolytic flux and glucose oxidation[76 77]; pentose cycle flux[78 79]; lactate output[77]; islet metabolites[65 67 77 80]; activities of islet enzymes[80-83]; islet ATP[63].

Further evidence for a close association of metabolism and insulin release has come from studies of the specificity of the secretory response to sugars and derivatives[62] and the ability of these agents to serve as fuels for the β-cells as shown by oxidation rate and maintenance of islet ATP concentration[63]. Of a large number of analogues of glucose tested on mouse and rat islets no non-metabolised sugar was found capable of initiating insulin release[63]. D-glyceraldehyde which induces action potentials in mouse islets[48] was found to initiate release of insulin and to be metabolised by mouse islets. Since glyceraldehyde enters the glycolytic pathway at the triose phosphate level, its metabolism unlike that of glucose is not inhibited by mannoheptulose[63]. The substrate-site model would predict therefore that glyceraldehyde-stimulated insulin release would not be inhibited by mannoheptulose. This prediction has been verified[63].

The interpretation of the effects of metabolites on insulin release is complicated by the partial dependence of islets on exogenous fuel for the maintenance of their ATP content. Insulin release is inhibited by agents which inhibit ATP synthesis, and the ATP content of mouse islets incubated without exogenous substrates shows a steady decline to about 50% of the initial value in 2 h[63]. Thus one action of a metabolised sugar may be to increase islet ATP to concentrations necessary for stimulation of insulin release to occur. Such an explanation for the specificity of secretory responses to sugars has not been entirely excluded, but seems unlikely in view of the lack of correlation of rates of insulin release and islet ATP content in the presence of various initiators and potentiators of release[63].

These studies are consistent with but do not establish the substrate-site hypothesis. However further studies of the specificity of the secretory response to sugars have broadened the scope of these considerations when it was found that in the presence of a substimulatory concentration of glucose (2.5–5 mM) a marked increase in insulin release was elicited by fructose, *N*-acetylglucosamine or (in rat but not mouse islets) galactose; glyceraldehyde and leucine can substitute for glucose. Mannoheptulose inhibited the potentiation of insulin release by *N*-acetylglucosamine with leucine or glyceraldehyde as initiators although it did not inhibit glyceraldehyde or leucine-stimulated release *per se*.

These results suggest that a single-site glucoreceptor model may be inadequate. The models depicted in Figure 2.1 have been proposed. The upper panel depicts a scheme retaining the essential features of the regulator-site model but incorporating initiating sites and potentiatory sites; binding to the former by glucose, mannose, glyceraldehyde and leucine leads to insulin release. Binding at the potentiator site does not elicit insulin release in the absence of binding at the initiator site, but, when the initiator site is activated, the presence of a potentiator, via co-operative interaction between the two units, leads to an increased rate of insulin release. Mannoheptulose inhibits the interaction of glucose and mannose but not leucine or glyceraldehyde with their initiator sites and is also capable of inhibiting the potentiator site. The modified form of the substrate-site hypothesis is shown in the lower panel. The essential feature here is that metabolism of the initiators leads to production of an intracellular agent triggering insulin release: it is the action of this metabolite that is potentiated by binding of a potentiator to the potentiator site. Mannoheptulose inhibits glucose metabolism and also the binding of sugars to the potentiator site. We cannot decide unequivocally between these models but at the present time the weight of evidence leads us to favour the second alternative.

Metabolite signal and insulin release

The key role assigned to a glucose metabolite by the substrate site hypothesis has prompted search for possible candidates. The stimulation of insulin release by pentitols led to suggestions that the pentose cycle may be of importance in initiating insulin release[84] but measurements of the flux through this pathway showed that only a small percentage of glucose is oxidised by this route: furthermore, as the glucose concentration is raised there is no change in this rate[78 79]. The level of hexose phosphates including G6P, FDP and 6PG is a possible candidate since parallel changes are seen with this parameter and insulin release under a variety of conditions[77 80 85]. However the possible importance of intermediates between triose phosphate and pyruvate has been stressed by Hellman[86]: an accumulation of FDP and intermediates above this metabolic stage was found when insulin secretion was inhibited by adrenaline or omission of Ca^{2+}. These data were interpreted as indicating that the sequence triosephosphate-dehydrogenase-phosphoglycerate kinase was a rate-limiting step in β-cell glycolysis of direct significance for regulation of insulin release.

Since glucose entry into β-cells is rapid[80] the rate-limiting step for glucose entry into β-cell metabolism is phosphorylation of the sugar. In mouse islets three enzymes have been detected that may control this rate[80 81]

These are hexokinase with a low K_m for glucose (0.1 mM), a glucokinase with a high K_m for glucose (10mM) and glucose-6-phosphatase. The properties of these enzymes have been shown to provide some basis for a gluco-receptor mechanism based on glucose phosphorylation[81].

Transduction of response to sugars

The interaction of glucose or a metabolite thereof with the β-cell appears, as described above, to result in a change in the cytosolic Ca^{2+} concentration. The mechanism of this transduction process is unknown but it may be relevant that islet glucose metabolism is influenced by the extracellular ionic environment. Glucose utilisation and oxidation is depressed in the absence of extracellular Na^+ [34 62 69]; this effect may be explicable in terms of a depression of glucose oxidation by a reduced ATP turnover in the absence of Na^+. Alternatively glucose entry into islets may be linked to Na^+ entry as proposed for intestinal mucosa. If this were the case then glucose utilisation by islets might lead to influx of Na^+ which may lead to an increased influx of Ca^{2+} by the mechanism suggested by Milner and Hales[87] and hence excite secretion. Glucose oxidation is also depressed in the absence of extracellular Ca^{2+} [69] but the relationship of this finding to release is not clear.

Hormones

The available data suggest that most if not all hormonal effects on insulin release may be mediated by changes in the β-cell cAMP content, according to the classical second messenger concept. Thus glucagon and ACTH have insulinotropic activity *in vivo* and *in vitro;* these effects are potentiatory, i.e. dependent on the presence of glucose or other initiator (for references see Malaisse[38]). Islet cAMP is increased by glucagon[49] and islet adenylcyclase is activated by glucagon[52 56] and by ACTH[55]. There is as yet no firm indication as to the physiological significance of these effects.

Considerable effort has been exerted to determine the nature of the intestinal factor(s), that possess(es) insulinotropic activity. The finding that oral glucose evokes more insulin release than a comparable loading with intravenous glucose[88] has been ascribed to the release of hormones from the intestinal tract which give an additional stimulus to insulin release after oral glucose. The term "enteroinsular axis" has been proposed to describe this concept[89]. Identification of the responsible intestinal factor requires satisfaction of two criteria: firstly, the demonstration of an increased release of the factor in response to oral glucose and, secondly, evidence that physiological concentrations of the factor modify the rate of insulin release in response to glucose. For references to the voluminous and often contradictory evidence relating to the effects of secretin, cholecystokinin–pancreozymin, gastrin and enteroglucagon on insulin release the reader is referred to several recent reviews[80][90][91]. Unequivocal evidence that any of these known gastrointestinal hormones may mediate the physiological response to oral as opposed to intravenous glucose is lacking. Two groups of workers have prepared extracts of intestine that elicited insulin release but were free of the then known gastrointestinal hormones[92][93]. However since these materials have not so far been characterised the criteria given above cannot be satisfied. Recently evidence has been adduced that gastric inhibitory peptide (GIP), a polypeptide identified and characterised as an inhibitor of gastric secretion, may be the major intestinal factor influencing the insulin response to oral glucose[124]. Physiological concentrations of highly purified GIP were infused intravenously into normal subjects. Administration of GIP alone did not increase blood insulin but GIP and glucose together greatly potentiated the rise in insulin level compared with glucose alone. The concentration of GIP used resulted in a serum GIP level similar to that achieved after ingestion of glucose. Effects of GIP on islet metabolism have not yet been described.

Adrenaline inhibits secretion in response to glucose, leucine, tolbutamide, ouabain, Ba^{2+}, high K^+, glucagon, theophylline and dbcAMP[94][95]~. The inhibition is abolished by α-adrenergic blocking agents but not by β-adrenergic blocking agents[43]. Adrenaline inhibits β-cell adenycyclase[52] and protein phosphokinase[53] and lowers β-cell cAMP content[49]. However the inhibition of release is not overcome by exogenous dibutyryl cAMP[96], suggesting that the inhibitory effects of adrenaline may not be ascribed solely to lowering of cAMP via α-adrenergic stimulation. Inhibition of glucose metabolism and accumulation of hexose phosphates in islets exposed to adrenaline has been found by Hellman[86], but the degree of inhibition of insulin release did not correlate well with the inhibition of glycolytic flux[97]. The inhibitory action, whatever its initial site of action, may be mediated by a decreased uptake of Ca^{2+} by the β-cell[98]: an inhibition of alkaline-earth cation transport would be compatible with failure of Ba^{2+} to overcome adrenaline inhibition of release[95]. Other modes of action cannot be excluded. The possible physiological significance of adrenaline inhibition of insulin release has been reviewed by Malaisse[38].

Insulin release induced by amino acid and other fuels
A number of amino acids whose relative potency varies from species to species stimulate insulin release under various conditions both *in vivo* and *in*

vitro[38][91]. Of these, leucine is an initiator of secretion whereas others are potentiators requiring the presence of glucose. Hellman and colleagues[99][100] studied the metabolism of amino acids in islets from obese hyperglycaemic mice. They showed that D-and L-leucine are transported into islets by the same carrier system but only L-leucine elicits insulin release. Alanine, arginine and leucine are concentrated by islets; alanine uptake is Na^+-dependent but leucine uptake is not. Omission of Ca^{2+} did not affect the uptake of leucine or of alanine. Leucine and alanine are oxidised but arginine is not; glucose stimulates the oxidation of alanine but depresses that of leucine, and, in this species, leucine and arginine, but not alanine, stimulate insulin release. Release is stimulated by the non-metabolised analogue of leucine, 2-aminobicyclo [2,2,1] heptane-2-carboxylic acid (BCH) but there is no direct evidence that BCH does not influence islet cell metabolism. It is difficult to draw firm conclusions at the present time about the mechanisms involved in the β-cell response to amino acids.

Secretory responses to other fuels such as fatty acids and ketone bodies have been observed[38][91]. Fatty acids have been shown to be oxidised by islets but little information is currently available to formulate suggestions as to the mode of action of these fuels. One would expect interactions between fat and carbohydrate metabolism but this has not been systematically investigated in the β-cell.

PHARMACOLOGICAL AGENTS

Sulphonylureas

Despite the long usage of sulphonylureas in diabetic therapy their mode of interaction with the β-cell has not been elucidated. It has been shown that sulphonylureas are restricted to the extracellular water in mouse islets[57], hence their presumed locus of interaction is the cell membrane. There is some evidence discussed above that this interaction may involve the adenylcyclase/phosphodiesterase system. However it is difficult to accept that sulphonylurea action is exerted solely through cAMP, since elevation of cAMP by caffeine does not evoke insulin release in the absence of glucose, whereas release evoked by sulphonylureas, although most marked in the presence of glucose is not absolutely dependent on glucose. Effects of sulphonylureas on ^{45}Ca-handling by the β-cell have been described by Malaisse *et al.*[95]

Diazoxide

CAMP has also been implicated in the inhibitory action of diazoxide on insulin release[49][53] but other sites of action are not excluded. Thus in the absence of extracellular Ca^{2+} diazoxide caused an immediate and sustained increased in ^{45}Ca efflux from perfused rat islets[101].

Membrane-active agents

Recent studies have shown that organic mercurials such as *p*-chloromercuribenzoate (PCMB) are potent stimulators of release in a number of secretory cell types[45][102][103] In contrast to other initiators of release, PCMB appears relatively insensitive to removal of extracellular Ca^{2+} in pituitary[103] and islets[102]. Other thiol reagents also stimulate insulin release. At low concentrations iodoacetamide stimulates release but, in contrast to PCMB, requires the presence of extracellular glucose[68]. These findings lead Hellman *et al.*[68][102] to suggest that relatively superficial thiol groups in the β-cell membrane may have a fundamental role in β-cell secretory responses.

CONTROL OF INSULIN BIOSYNTHESIS[8]

Glucose stimulates the biosynthesis of proinsulin with a concentration dependence similar to glucose-evoked release[104]. Both processes are inhibited by mannoheptulose[105] and both may be potentiated by caffeine[106]. These data suggest that the two processes may share common control characteristics. However the two processes are not obligatorily coupled: stimulation of release can occur without stimulation of synthesis, e.g. in the presence of cycloheximide[107] and glucose stimulates proinsulin biosynthesis in the absence of extracellular Ca^{2+} when release is not stimulated[8][106][108]. The latter finding suggests also that the major effect of Ca^{2+} on islet function is on a terminal event in the release mechanism. On the other hand at low pH or in the absence of extracellular Na^+ there is a parallel reduction in glucose-induced insulin biosynthesis, Ca^{2+} uptake, and insulin release[108], suggesting therefore that the β-cell glucose-recognition process involves a Na^+ dependent step.

Figure 2.2 shows a working model summarising our conclusions about the regulation of insulin release. Initiators of release such as glucose may increase cytosol calcium concentration and potentiators such as hormones may increase cytosol cAMP concentration through effects on a calcium transporter or adenylcyclase (input transducers). Ca^{2+} and cAMP act as intracellular transmitters which activate the output system (components involved in exocytosis) through output transducers (a postulated calcium-binding regulator and cAMP-dependent protein kinase).

INSULIN SECRETION IN DIABETES AND OTHER RELEVANT CLINICAL DISORDERS

Plasma insulin concentration in diabetes

The classical studies of plasma insulin concentration in diabetes, carried out by Berson and Yalow following their discovery of radioimmunoassay, showed that plasma insulin concentrations in insulin-requiring diabetics are subnormal and that the plasma insulin response to glucose or tolbutamide is blunted or absent[109]. In this form of diabetes the degree of impairment of glucose tolerance may correlate with the plasma insulin response. These findings in conjunction with the diminished number of islets and low insulin content of the pancreas are consistent with severe loss of β-cell function as the basis of the disorder. In maturity-onset diabetes the plasma insulin response to glucose is delayed and may lead ultimately to hyperinsulinaemia by comparison with euglycaemic control[109 110]. More stringent investigations by Perly and Kipnis[111] made allowance for the effects of obesity and the degree of hyperglycaemia in assessing the plasma insulin response to glucose in maturity-diabetics. When allowance was made for these factors, the results showed that the plasma insulin response at equivalent plasma glucose concentrations is impaired in the maturity-onset diabetic, irrespective of whether obesity is present or absent. Obesity in normoglycaemic controls or diabetics leads to an exaggerated plasma insulin response to glucose. In the insulin-requiring diabetic, insulin deficiency would appear to be responsible for glucose intolerance. In the maturity—onset diabetic, insulin insufficiency and insulin insensitivity may both contribute.

The scope of these considerations was widened by the observation that the plasma insulin response to intravenous glucose loading may be impaired in the non-diabetic monozygotic twin siblings of diabetics and in 15—20% of adults or children with normal glucose tolerance. These findings led to the suggestion that an impaired plasma—insulin response to glucose may be the result of an inherited abnormality in the β-cell which sets the stage for the subsequent development of diabetes[112]. Comparisons of the plasma-insulin response to glucose in portal-vein and peripheral blood have supported the suggestion that the impaired plasma—insulin response reflects an impaired insulin secretory response[113]. The observations and the conclusions drawn from them have been criticised on a number of grounds. In particular it has been noted that information on the incidence of low responders in non-diabetics is based on a restricted population thus allowing no firm conclusions to be drawn about inheritance. Moreover the

(a)

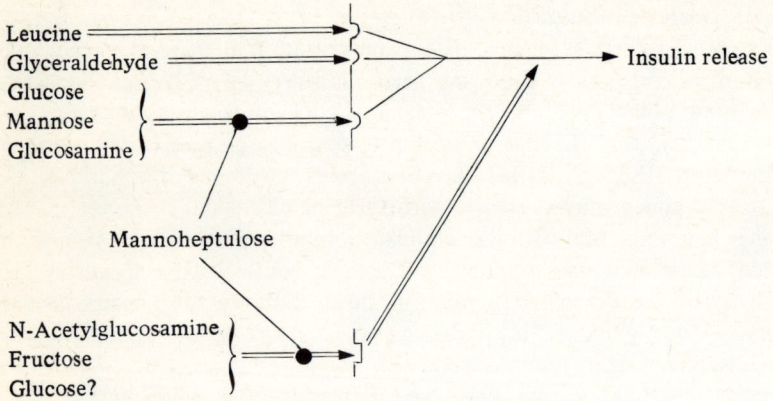

Mannoheptulose

(b)

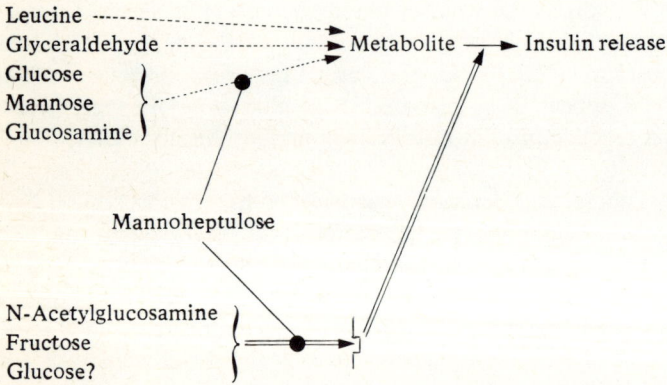

Mannoheptulose

Figure 2.1 Glucoreceptor models (a) The β-cell is depicted as possessing initiator sites
(?) capable of eliciting insulin release when activated and a potentiator site (?), activation
of which potentiates the effects of activation of the initiator site. In (b) the primary
stimulators of insulin release act through the formation of a trigger metabolite whose
action may be potentiated by activation of the potentiator site. The inhibitory affects of
mannoheptulose are shown as

INPUT SYSTEMS

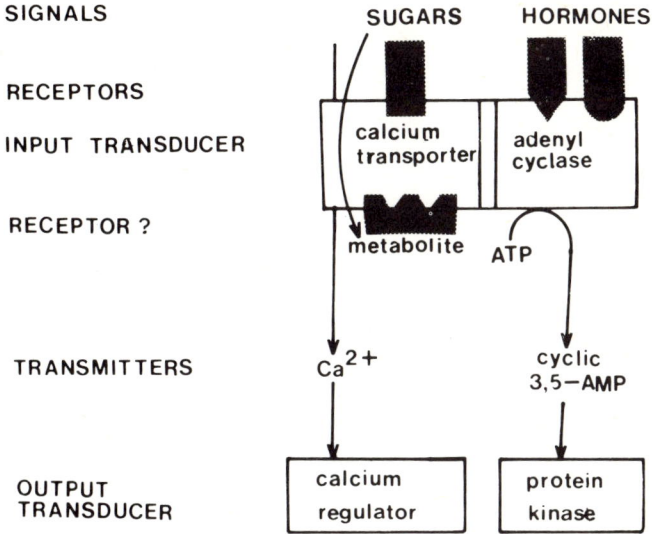

OUTPUT SYSTEMS

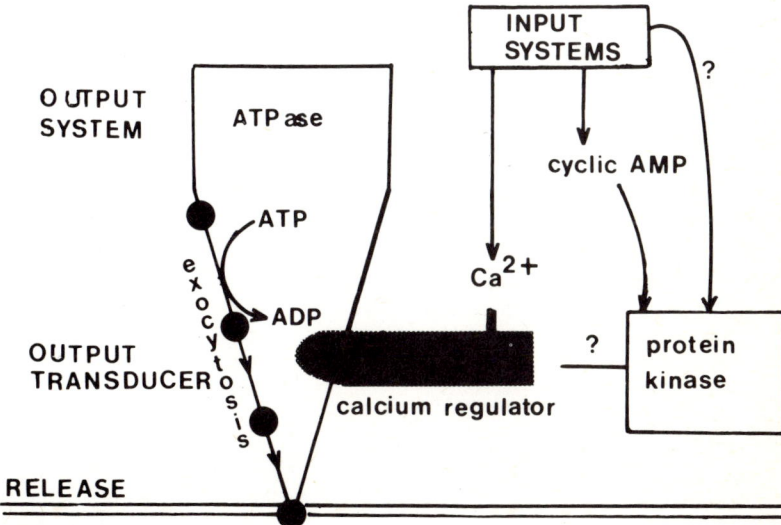

Figure 2.2

studies by Tattersall and Pyke[114] on diabetes in monozygotic twins has indicated that concordance is more frequent in diabetes developing over the age of 40 than in diabetes developing before that age. This has suggested that factors other than genetic predisposition may be of greater importance in the aetiology of insulin-requiring diabetes in young people.

Cerasi *et al.*[115] have provided quantitative information on the relationship between blood glucose and plasma insulin in diabetics, and in non-diabetics exhibiting good or poor plasma–insulin responses to intravenous glucose. The poor responders were described as pre-diabetic but this description lacked supporting evidence. The results showed that the curve relating plasma insulin concentration to blood glucose concentration is shifted to the right in the diabetics and the non-diabetics exhibiting a poor response. The K_m for glucose-induced insulin secretion may therefore be increased but the data did not permit conclusions regarding the maximum rate of insulin release. The studies were prompted by a number of reports of high plasma insulin response to oral glucose in potential diabetics and patients with mild carbohydrate intolerance[116-118].

In terms of the models for insulin secretion considered in Figures 2.1 and 2.2 there are a number of different possible explanations for a blunted insulin secretory response to glucose in diabetics of 'potential diabetics'. The population of β-cells may be diminished; the glucoreceptor may be modified so that sensitivity to stimulation by glucose is reduced, or alternatively the input transducer may be modified; amplification by cAMP may be diminished through defects in the hormone receptors or adenylcyclase or in cAMP phosphodiesterase; lastly there may be defects in the output system such that an otherwise effective signal results in defective output of insulin. A crucial question is whether the impaired

---

Figure 2.2 model for the regulation of insulin release The β-cell is depicted as possessing input systems (upper panel) which recognise and respond to stimuli by increasing the cytosolic concentration of Ca^{2+} and in some cases also of cAMP. The signal of an increased glucose concentration is detected by a receptor which may be sensitive to glucose itself and/or to a metabolite of glucose: this information is translated into a rise in Ca^{2+} concentration through effects on a Ca^{2+} transporter. Hormonal signals are exerted via adenylcyclase. The output system (lower panel) comprises those components involved in exocytosis, and the rise in Ca^{2+} activates this system by combination with a Ca^{2+}-sensitive regulator of the output system. The properties of the Ca^{2+} regulator may be modified by a cAMP–dependent protein kinase, thus permitting modulation of the secretory response to intiators such as sugars by parallel cAMP content

plasma–insulin response in diabetics is specific for glucose. Evidence on this point is not comprehensive. There is evidence that diabetics or 'potential diabetics' may show normal plasma–insulin responses to glucagon[122] or tolbutamide, although evidence for tolbutamide is conflicting. These observations have led Cerasi and Luft to conclude that the defect is specific for glucose and is located in the glucoreceptor mechanism. Floyd *et al*[119]. have observed a diminished plasma–insulin response in diabetics to the intravenous administration of arginine, of a mixture of ten essential amino acids and to a protein meal.

One of the major difficulties in drawing firm conclusions from clinical studies is the lack of information which can only be obtained *in vitro* about the insulin secretory responses of the human islet. Human islets which exhibit secretory and metabolic responses have been prepared[120] but detailed information about which agents are initiators and which are potentiators is lacking. In the mouse islet a number of secretory responses may be dependent on the presence of an initiator such as glucose. These may include secretory responses to caffeine, theophylline or isobutyl methylxanthine, to hormones such as glucagon, to some amino acids (e.g. arginine), but not to others (e.g. leucine). If the initiator response to glucose is lost (e.g. with mannoheptulose) then the potentiator response (e.g. with caffeine or glucagon or arginine) is also lost. If the human islet shows the same pattern as the mouse islet then the diminished response to some amino acids (e.g. arginine) could be the result of a diminished effect of glucose as an initiator. The apparently normal plasma insulin response to glucagon would then offer difficulties in interpretation. In our view lack of information about the pattern of insulin secretory responses in the human islet precludes firm interpretation of data obtained in patients.

Cyclic AMP and insulin secretory responses in diabetes

Cerasi and Luft[121] have suggested that a low islet concentration of cAMP is responsible for the impaired insulin secretory responses in diabetes. Evidence for this view is based principally on the observation that theophylline, which may elevate islet cAMP by inhibiting the cyclic nucleotide phosphodiesterase, can restor towards normal the plasma insulin response to glucose infusion of maturity-onset diabetics. They have suggested that the defect is a specific failure of glucose to acutely activate adenylcyclase and to elevate islet cAMP[112]. For reasons which have been summarised previously the bulk of experimental evidence does not support the view that glucose initiates insulin release by elevating islet cAMP concentration. There is however compelling evidence that glucose may, by indirect and long-term effects, modulate islet adenylcylase activity and islet cAMP concentration (see below).

Starvation diabetes

Several days of starvation or of low carbohydrate diet can lead in man to impaired glucose tolerance and to a delayed plasma–insulin response to glucose[110] [122]. In rats, fasting may also lead to an impaired plasma–insulin response to glucose, and islets from starved rats may show an impaired insulin secretory response to glucose when compared with islets from fed rats. The plasma–insulin responses to theophylline were, however, comparable in fed and starved rats. These changes show some similarity to those seen in maturity-onset diabetes.

The possible role of adenylcyclase in these changes in insulin secretion during starvation have been investigated by Howell *et al.*[123] The insulin secretory response to glucose was diminished in islets from starved rats but restored to normal by addition of theophylline. The activity of adenylcyclase (basal or glucagon-stimulated) was markedly reduced in homogenates of islets from starved rats when compared with fed controls. Conversely glucose loading (by injection over 5 h or feeding over 5 days) increased islet adenylcyclase activity. Incubation of islets *in vitro* in the presence of glucose for periods in excess of one hour led to an increase in adenylcyclase activity which was progressive for up to seven hours. These observations indicate that glucose may exert long-term effects on the activity of adenylcyclase. The mechanism is not known but it does not appear to involve formation of adenylcyclase by protein synthesis. These findings provide the basis for an attractive hypothesis in which availability of glucose may modulate insulin secretory responses to hyperglycaemia by long-term control of the activity of adenycyclase and hence of islet cAMP concentrations. It seems possible that these observations could provide a mechanism whose disturbance in diabetes may lead to an impaired insulin secretory response. However, the apparently normal insulin secretory response to glucagon in diabetes is difficult to reconcile with this type of mechanism, and so is the increase in secretory response induced by preloading with glucose in maturity onset diabetes.

References

1. Lacy, P. E. and Greider, M. H. (1972). Ultrastructural organisation of mammalian pancreatic islet. *Handbook of Physiology,* **Section 7, Vol. 1,** 77
2. Lacy, P. E. (1967). The pancreatic beta cell. Structure and Function. *N. E. J. Med.,* **276,** 187
3. Orci, L., Stauffacher. W., Beaven, D., Lambert, A. E., Renold, A. E. and Rouitler, C. (1969). Ultrastructural events associated with the action of tolbutamide and

glybenclamide on pancreatic β-cells *in vivo* and *in vitro*. *Acta Diabet. Cat.*, **6, (Suppl. 1),** 271

4. Orci, L., Amherdt, M., Kanazawa, Y., Lambert, A. E. and Stauffacher, W. (1972). Morphological events in insulin secreting cell. In: *Hormones Pancreatiques, Hormones de l'Eau et des Electrolytes,* 15 (Boulogne, France: INSERM)

5. Steiner, D. F. and Oyer, P. B. (1967). The biosynthesis of insulin and a probable precursor of insulin by a human islet cell adenoma. *Proc. Nat. Acad. Sci.,* **57,** 473

6. Kemmler, W., Peterson, J. D., Rubenstein, A. H. and Steiner, D. F. (1972). On the biosynthesis, intracellular transport and mechanism of conversion of proinsulin to insulin and C-peptide. *Diabetes,* **21, (Suppl. 2),** 572

7. Howell, S. L., Kostianovsky, M. and Lacy, P. E. (1969). Beta granule formation in isolated islets of Langerhans. A study by electron microscope radioautography. *J. Cell. Biol.,***42,** 695

8. Steiner, D. F., Kemmler, W., Clark, J. L., Oyer, P. E. and Rubenstein, A. H. (1972. The biosynthesis of insulin. *Handbook of Physiology,* **Section 7, Vol. 1,** 175

9. Coore, H. G., Hellman, B., Phil, E. and Taljedal, I. B. (1969). Physiochemical characterisation of insulin secretion granules. *Biochem. J.,* **111,** 107

10. Howell, S. L., Fink, C. J. and Lacy, P. E. (1969). Isolation and properties of secretory granules from rat islets of Langerhans. I. Isolation of a secretory granule fraction. *J. Cell. Biol.,* **41,** 154

11. Howell, S. L., Young, D. A. and Lacy, P. E. (1969). Isolation and properties of secretory granules from rat islets of Langerhans. III. Studies of the stability of the isolated beta granules. *J. cell. Biol.,* **41,** 167

12. Lacy, P. E. (1961). Electron microscopy of the beta cell of the pancreas. *Amer. J. Med.,* **31,** 851

13. Findlay, J. A., Gill, J. R., Irvine, G., Lever, J. D. and Randle, P. J. (1968). Cytology of β-cells, in rabbit pancreas pieces incubated *in vitro*. Effects of glucose and tolbutamide. *Diabetologia,* **4,** 150

14. Lacy, P. E. (1970). Functional morphology of the islet cells. *Nobel Symp.,* **13,** 109

15. Orci, L., Lambert, A. E., Amherdt, M., Cameron, D., Kanazawa, Y. and Stauffacher, W. (1970). The autonomous nervous system and the β-cell: metabolic and morphological observations made in spiny mice *(Acomys cahirinus)* and in cultured fetal rat pancreas. In: On the pathogenesis of diabetes mellitus. (R. Luft and P. J. Randle, editors) *Acta Diabet. Lat.,* **7, (Suppl. 1),** 184

16. Creutzfeldt, W., Creutzfeldt, C. and Frerichs, H. (1970). Evidence for different modes of insulin secretion. In: *The Structure and Metabolism of the Pancreatic Islets,* 181 (S. Falkmer, B. Hellman and I-B. Taljedal, editors) (Oxford: Pergamon Press)

17. Orci, L. (1973). Exocytosis-endocytosis coupling in the pancreatic beta cell. *Science,* **181,** 561

18. Matthews, E. K. (1970). Calcium and hormone release. In: *Calcium and Cellular Function,* 163 (A. W. Cuthbert, editor) (London: Macmillan)

19. Olmsted, J. B. and Borisy, G. G. (1973). Microtubules. *Ann. Rev. Biochem.,* **42,** 507

20. Shelanski, M. J. (1973). Chemistry of the filaments and tubules of brain. *J. Histol. Cytol.,* **21,** 529

21. Weisenberg, R. C. (1972). Microtubule formation *in vitro* in solutions containing low calcium concentrations. *Science,* **177,** 1104

22. Goodman, D. B. P., Rasmussen, H., DiBella, F. and Guthrow, C. E. (1970). Cyclic adenosine 3′,5′-monophophate-stimulated phosphorylation of isolated neurotubule subunits. *Proc. Nat. Acad. Sci.*, **67**, 652

23. Malaisse, W. J. (1973). Insulin secretion: multifactorial regulation for a single process of release. *Diabetologia, **9**, 167

24. Malaisse, W. J., Malaisse-Lagae, F., Walker, M. O. and Lacy, P. E. (1971). The stimulus-secretion coupling of glucose-induced insulin release. V. The participation of a microtubule-microfilamentous system.

25. Devis, G., Obberghen, E., Somers, G., Malaisse-Lagae, F., Orci, L. and Malaisse, W. J. (1974). Dynamics of insulin release and microtubular-microfilamentous system. II. Effect of vincristine. *Diabetologia, **10**, 53

26. Mizel and Wilson (1972). Effect of colchicine on nucleoside transport of mammalin cells. *Biochemistry*, **11**, 2573

27. Wilson, L., Bryan, J., Ruby. A. and Mazia, D. (1970). Precipitation of proteins by vinblastine and calcium ions. *Proc. Nat. Acad. Sci.*, **66**, 807

28. Kaminer, B. and Kimura, J. (1972). *Science*, **176**, 406

29. Wessels, N. K., Spooner, B. S., Ash, J. F., Bradley, M., Luduena, M. A., Taylor, E. L., Wrenn, J. T. and Yamaela, K. M. (1971). Microfilaments in cellular developmental processes. *Science*, **171**, 135

30. Orci, L., Gabbay, K. H. and Malaisse, W. J. (1972). Pancreatic beta-cell web: its possible role in insulin secretion. *Science*, **175**, 1128

31. Malaisse, W. J., Hager D. L. and Orci, L. (1972). The stimulus–secretion coupling of glucose-induced insulin release IX. The participation of the beta cell web. *Diabetes*, **21, (Suppl. 2)**, 594

32. Spudich, J. A. and Lin, S. (1972). Cytochalasin B, its interaction with actin and actomyosin from muscle. *Proc. Nat Acad. Sci.*, **69**, 442

33. Estenson, R. D. and Plagemann, P. G. W. (1972). Cytochalasin B: inhibition of glucose and glucosamine transport. *Proc. Nat. Acad. Sci.*, **69**, 1430

34. Matschinsky, F. M. and Ellermann, J. (1973). Dissociation of the insulin releasing and the metabolic function of hexoses in islets of Langerhans. *Biochem. Biophys. Res. Comm.*, **50**, 193

35. Schauder, P. and Frerichs, H. (1974). Cytochalasin B: inhibition of glucose-induced insulin release from isolated rat pancreatic islets. *Diabetologia, **10**, 85

36. Berl, S., Puszrin, S. and Nicklas, W. J. (1973). Actomyosin-like protein in brain. *Science*, **179**, 441

37. Randle, P. J. and Hales, C. N. (1972). Insulin release mechanisms. *Handbook of Physiology*, **Section 7, Vol. 1**, 219

38. Malaisse, W. J. (1972). Hormonal and environmental modification of islet activity. *Ibid*, 237

39. Grodsky, G. M. (1970). Insulin and the pancreas. *Vitamins and Hormones*, **28**, 37

40. Grodsky, G. M. (1972). A threshold distribution hypothesis for packet storage of insulin. II Effect of Calcium. *Diabetes*, **21, (Suppl. 2)**, 584

41. Grodsky, G. M. (1972). A threshold distribution hypothesis for packet storage of insulin and its mathematical modelling. *J. Clin. Invest.*, **51**, 2047

42. Curry, D. L., Bennett, L. L. and Grodsky, G. M. (1968). Requirement for calcium ion in insulin secretion by the perfused rat pancreas. *Amer. J. Physiol.*, **214**, 174

43. Hales, C. N. (1970). Ion fluxes and membrane function in β-cells and adipocytes. *Acta Diabet. Lat.*, **7, (Suppl. 1)**, 64

44. Hales, C. N. and Milner, R. D. G. (1968). Cations and the secretion of insulin from rabbit pancreas *in vitro. J. Physiol.,* **199,** 177

45. Dormer, R. L., Kerby, A. L., McPherson, M., Manley, S. Ashcroft, S. J. H., Schofield, J. G. and Randle, P. J. (1974). The effects of nickel of secretory systems. Studies on the release of amylase, insulin and growth hormone. *Biochem. J.,* **140,** 135

46. Dean, P. M. and Matthews, E. K. (1970). Glucose-induced electrical activity in pancreatic islet cells. *J. Physiol.,* **210,** 255

47. Dean, P. M. and Matthews, E. K. (1970). Electrical activity in pancreatic islet cells; effect of ions. *J. Physiol.,* **210,** 265

48. Dean, P. M. and Matthews, E. K. (1972). The bioelectrical properties of pancreatic cells: effect of diabetogenic agents. *Diabetologia,* **8,** 173

49. Montague, W. and Cook, J. R. (1971). The role of adenosine 3′,5′-cyclic monophosphate in the regulation of insulin release by isolated rat islets of Langerhans. *Biochem. J.,* **122,** 115

50. Cooper, R. H., Ashcroft, S. J. H. and Randle, P. J. (1973). Concentration of adenosine 3′,5′-cyclic monophosphate in mouse pancreatic islets measured by a protein-binding radioassay. *Biochem. J.,* **134,** 599

51. Ashcroft, S. J. H., Randle, P. J. and Taljedal, I-B. (1972). Cyclic nucleotide phosphodiesterase activity in normal mouse pancreatic islets. *Febs Letters,* **20,** 263

52. Howell, S. L. and Montague, W. (1973). Adenylate cyclase activity in isolated rat islets of Langerhans. Effects of agents which alter rates of insulin secretion. *Biochem. Biophys. Acta,* **320,** 44

53. Montague, W. and Howell, S. L. (1973). The mode of action of adenosine 3′,5′-cyclic monophosphate in mammalian islets of Langerhans. Effects of insulin secretagogues on islet cell protein kinase activity. *Biochem. J.,* **134,** 321

54. Charles, M. A., Fanska, R., Schmid, F. G., Forsham, P. H. and Grodsky, G. M. (1973). Adenosine 3′,5′-monophosphate in pancreatic islets: glucose-induced insulin release. *Science,* **179,** 569

55. Grill, V. and Cerasi, E. (1973). Activation by glucose of adenyl cyclase in pancreatic islets of the rat. *Febs Letter,* **33,** 311

56. Kuo, W. N., Hodgin, W. S. and Kuo, J. F. (1973). Adenylate cyclase in islets of Langerhans. Isolation of islets and regulation of adenylate cyclase activity by various hormones and agents. *J. Biol. Chem.,* **248,** 2705

57. Hellman, B., Sehlin, J. and Taljedal, I-B. (1971). The pancreatic β-cell recognition of insulin secretagogues. II. Site of action of tolbutamide. *Biochem. Biophys. Res. Comm.,* **45,** 1384

58. Atkins, T. and Matty, A. J. (1971). Adenyl cyclase phosphodiesterase activity in isolated islets of Langerhans of obese mice and their lean litter-mates: the effect of glucose, adrenaline, and drugs on adenyl cyclase activity. *J. Endocrinol.,* **51,** 67

59. Sams, D. J. and Montague, W. (1972). The role of adenosine 3′, 5′-cyclic monophosphate in the regulation of insulin release. Properties of islet cell adenosine 3′,5′-cyclic monophosphate phosphodiesterase. *Biochem. J.,* **129,** 954

60. Montague, W. and Howell, S. L. (1972). The mode of action of adenosine 3′,5′-cyclic monophosphate in mammalian islets of Langerhans. Preparation and properties of islet-cell protein phosphokinase. *Biochem. J.,* **129,** 551

61. Rasmussen, H. (1970). Cell communication, calcium ion and cyclic adenosine monophosphate. *Science,* **170,** 404

62. Ashcroft, S. J. H., Bassett, J. M. and Randle, P. J. (1972). Insulin secretion mechanisms and glucose metabolism in isolated islets. *Diabetes*, **21, (Suppl.2),** 538

67. Ashcroft, S. J. H., Capito, K. and Hedeskov, C. J. (1973). Time course studies of glucose-induced changes in glucose-6-phosphate and fructose 1:6 diphosphate content of mouse and rat pancreatic islets. *Diabetologia*, **9,** 299|

68. Hellman, B., Idahl, L.A., Lernmark, Å., Sehlin, J. and Taljedal I-B. (1973) Iodoacetamide-induced sensitisation of the pancreatic β-cells to glucose stimulation. *Biochem. J.*, **132,** 775

69. Hellman, B., Idahl, L. Å., Lernmark, Å., Sehlin, J. and Taljedal, I-B. (1974). The pancreatic β-cell recognition of insulin secretagogues. Effects of calcium and sodium on glucose metabolism and insulin release.

70. Landgraf, R., Kotler-Brajtburg, J. and Matschinsky, F. M. (1971). Kinetics of insulin release from the perfused rat pancreas caused by glucose, glucosamine and galactose, *Proc. Nat. Acad. Sci.*, **68,** 536

71. Grodsky, G. M., Batts, A., Bennett, L. L., Vcella, C., Williams, N. B. and Smith, D. F. (1963). Effects of carbohydrates on secretion of insulin from isolated rat pancreas. *Amer. J. Physiol.*, **205,** 638

72. Coore, H. G. and Randle P. J. (1964). Regulation of insulin secretion studied with pieces of rabbit pancreas incubated *in vitro*. *Biochem. J.*, **93,** 66

73. Malaisse, W. (1969). *Etude de la Secretion Insulinique* in Vitro (Brussels: Editions Arscia)

74. Lacy, P. E., Young, D. A. and Fink, C. J. (1968). Studies on insulin secretion *in vitro* from isolated islets of the rat pancreas. *Endocrinology*, **83,** 1155

75. Price, S. (1973). Pancreatic islet cell membranes: extraction of a possible glucoreceptor. *Biochem. Biophys. Acta*, **318,** 459

76. Hellerstrom, C. (1967). Effects of carbohydrates on the oxygen consumption of isolated pancreatic islets of mice. *Endocrinology*, **81,** 105

77. Ashcroft, S. J. H., Hedeskov, C. J. and Randle, P. J. (1970). Glucose metabolism in mouse pancreatic islets. *Biochem. J.*, **118,** 143

78. Ashcroft, S. J. H., Weerasinghe, L. C. C. and Randle, P. J. (1972). The pentose cycle and insulin release in mouse pancreatic islets *Biochem. J.*, **126,** 525

79. Snyder, P. J., Kashket, S. and O'Sullivan, J. B. (1970). Pentose cycle in isolated islets during glucose-stimulated insulin release. *Amer. J. Physiol.*, **219.** 876

80. Matschinsky, F. M. and Ellerman, J. E. (1968). Metabolism of glucose in the islets of Langerhans. *J. Biol. Chem.*, **243,** 2730

81. Ashcroft, S. J. H. and Randle, P. J. (1970). Enzymes of glucose metabolism in normal mouse pancreatic islets. *Biochem. J.*, **119,** 5

82. Matschinsky, F. M., Kauffman, F. C. and Ellerman, J. E. (1968). Effect of hyperglycaemia on the hexosemonophosphate shunt in islets of Langerhans. *Diabetes*, **17,** 475

83. Brolin, S. E., Berne, M. K. and Linde,B. (1967). Measurement of the enzymatic activities required for ATP formation by glycolysis in the pancreatic islets of hyperglycaemic mice. *Diabetes*, **16,** 21

84. Montague, W. and Taylor K. W. (1968). Pentitols and insulin release by isolated rat islets of Langerhans. *Biochem. J.*, **109,** 333

85. Montague, W. and Taylor, K. W. (1967). Islet cell metabolism during insulin release. Effects of glucose, citrate, octanoate, tolbutamide, glucagon and theophylline- *Biochem. J.*, **115,** 257

86. Hellman, B. (1970). Methodological approaches to studies on the pancreatic islets. *Diabetologia*, **6**, 110

87. Milner, R. D. G. and Hales, C. N. (1970). Ionic mechanisms in the regulation of insulin secretion. In: *The Structure and Metabolism of the Pancreatic Islets*, 489 (S. Falkner, B. Hellman and I-B. Taljedal, editors) (Oxford: Pergamon Press)

88. McIntyre, N., Holdsworth, C. D. and Turner, D. S. (1964). New interpretation of oral glucose tolerance. *Lancet*, **ii**, 20

89. Unger, R. H. and Eisentraut, A. M. (1969). Entero-insular axis. *Arch. Int. Med.*, **123**, 261

90. Fajans. S. S. and Floyd, J. C. (1972). Stimulation of islet cell secretion by nutrients and by gastrointestinal hormones released during digestion. *Handbook of Physiology*, **Section 7, Vol. 1**, 473

91. Mirsky, I. A. (1973). Regulation of (insulin) secretion. In: *Methods in Investigative and Diagnostic Endocrinology*, 1238, **Vol. 2B** (S. A. Berson and S. Yalow, editors) (North Holland Publishing Co.)

92. Moody, A. J., Markussen, J., Fries, A. S., Stenstrup, C., Sundby, F., Malaisse, W. and Malaisse-Lagaw, F. (1970). The insulin releasing activities of extracts of pork intestine. *Diabetologia*, **6**, 135

93. Turner, D. S. and Marks, V. (1972). Enhancement of glucose stimulated insulin release by an intestinal polypeptide in rats. *Lancet*, **i**, 1095

94. Milner, R. D. G. and Hales, C. N. (1969). The interaction of various inhibitors and stimuli of insulin release studied with rabbit pancreas *in vitro*. *Biochem. J.*, **113**, 473

95. Malaisse, W. *et al.* (1972). Transport of L-leucine and D-leucine into pancreatic β-cells with reference to the mechanism of amino-acid-induced insulin release. *Biochim. Biophys. Acta* **266**, 436

96. Malaisse, W. J., Brisson, G. and Malaisse-Lagae, F. (1970). The stimulus-secretion coupling of glucose-induced insulin release. I. Interaction of epinephrine and alkaline earth cations. *J. Lab. Clin. Med.*, **76**, 895

97. Hellman, B., Sehlin, J. and Taljedal, I-B. (1970). *Med. Exp.*, **19**, 351

98. Malaisse, W., Malaisse, W. and Malaisse-Lagae, I. (1970). Uptake of ^{45}Calcium by isolated islets of Langerhans. *Diabetes*, **19**, 365

99. Hellman, B., Sehlin, J. and Taljedal, I-B. (1971). Uptake of alanine, arginine and leucine by mammalian pancreatic β-cells. *Endocrinology*, **89**, 1432

100. Hellman, B., Sehlin, J. and Taljedal, I-B. (1971). Effects of glucose and other modifiers of insulin release on the oxidative metabolism of amino acids on micro-dissected pancreatic islets. *Biochem. J.*, **123**, 513

101. Malaisse, W., Pipeleers, D. G. and Mahy, M. (1973). The stimulus-secretion coupling of glucose-induced insulin release. XII. Effects of diazoxide and gliclazide upon ^{45}Ca efflux from perifused islets. *Diabetologia*, **9**, 1

102. Bloom, G. D., Hellman B., Idahl, L-Å., Lernmark, Å, Sehlin, J. and Taljedal,I-B. (1972). Effects of organic mercurials on mammalian pancreatic β-cells. *Biochem. J.*, **129**, 241

103. Schofield, J. G. (1971). Effect of sulphydryl reagents on the release of ox growth hormone *in vitro*. *Biochim. Biophys. Acta*,**252**, 516

104. Lin, B. J., Nagy, B. R. and Haist, R. E. (1972). Effect of various concentrations of glucose on insulin biosynthesis. *Endocrinology*, **91**. 309

105. Lin, B. J. and Haist, R. E. (1969). Insulin biosynthesis: effect of carbohydrates and related compounds. *Can. J. Physiol. Pharmacol.*, **47**, 791

106. Lin, B. J. and Haist, R. E. (1973). Effects of some modifiers of insulin secretion on insulin biosynthesis. *Endocrinology, 92,* 735
107. Morris, G. E. and Korner, A. (1970). The effect of glucose on insulin biosynthesis by islets of Langerhans of the rat. *Biochim. Biophys. Acta,208,* 404
108. Pipeleers, D. G., Marichal, M. and Malaisse W. J. (1973). The stimulus-secretion coupling of glucose-induced insulin release. XIV. Glucose regulation of insular bio-synthetic activity. *Endocrinology, 93,* 1001
109. Berson, S. A. and Yalow, R. S. (1962). Immunoassay of plasma insulin. *Ciba Foundation Colloquia Endocrinol.,14,* 182–201
110. Hales, C. N. and Randle, P. J. (1963). Effects of low-carbohydrate diet and diabetes mellitus on plasma concentrations of glucose, non-esterified fatty acid and insulin during oral glucose-tolerance tests. *Lancet,* **i,** 790
111. Perly, M. and Kipnis, D. M. (1967). Plasma insulin responses to oral and intravenous glucose: studies in normal and diabetic subjects. *J. Clin. Invest.,46,* 1954
112. Cerasi, E. and Luft, R. (1973) Pathogenesis of genetic diabetes mellitus: further development of a hypothesis. *Mount Sinai J. Med.,40,* 334-349
113. Cerasi, E., Hallberg, D. and Luft, R. (1970). Simultaneous determination of insulin in brachial and portal veins during glucose infusion in normal and prediabetic subjects. *Horm. Metab. Res., 2,* 303
114. Tattersall, R. B. and Pyke, D. A. (1972). Diabetes in identical twins. *Lancet,* **ii,**1120
115. Cerasi, E., Ependic, S. and Luft, R. (1973). Dose-response relation between plasma-insulin and blood-glucose levels during oral glucose loads in prediabetic and diabetic subjects. *Lancet,* **i,** 794
116. Paulsen, E. P., Richenderfer, L. and Ginsberg-Fellner, F. (1968). Plasma glucose, free-fatty acids and immunoreactive insulin in sixty-six obese children. *Diabetes,* **17,** 261
117. Jackson, W. P. U., van Mieahe, E. and Keller, P. (1972). Insulin excess as the initial lesion in diabetes. *Lancet,* **i,** 1040
118. Reaven, G. M., Shen, S. W., Silvers, A. and Farquhar, J. W. (1971). Is there a delay in the plasma insulin response of patients with chemical diabetes mellitus? *Diabetes,* **20,** 416
119. Floyd, J. C. Jr, Fayans, S. S., Conn, J. W., Pek, S. and Knopf, R. F. (1970). In: *Early Diabetes* (R. Camerini-Davalos and H. S. Coles, editors) (New York and London: Academic Press)
120. Ashcroft, S. J. H., Bassett, J. M. and Randle, P. J. (1971). Isolation of human pancreatic islets capable of releasing insulin and metabolising glucose *in vitro. Lancet,* **i,** 888
121. Cerasi, E. and Luft, R. (1970). *Nobel Symp.,* **13,** 17–40
122. Cahill, G. F., Herrera, M. G., Morgan, A. P., Soeldner, J. S., Steinke, J., Levey, P. L., Richard, G. N. and Kipnis, D. M. (1966). Hormone fuel interrelationships during fasting. *J. Clin. Invest., 45,* 1751–1769
123. Howell, S. L., Green, I. C. and Montague W. (1973). A possible role of adenyl cyclase in the long-term dietary regulation of insulin secretion from rat islets of Langerhans. *Biochem. J., 136,* 343–349
124. Dupré, J., Ross, S. A., Watson, D. and Brown, J. C. (1973). Stimulation of insulin secretion by gastric inhibitory peptide in man. *J. Clin. Endocrinol. Metab., 37,* 826

3
THE GASTROINTESTINAL TRACT AND DIABETES MELLITUS

K. D. Buchanan

The gastrointestinal tract can be considered as the doorway through which all nutrients and energy sources of the body pass before they are utilised. Occasionally this doorway may be by-passed by parenteral feeding but this is not relevant in the normal individual. The passage of food through the doorway is not simply a passive event as the gut has numerous functions in preparing and processing the food for the body in addition to disposing of waste materials. The gastrointestinal tract acts as a mill, then digests the food before finally absorbing it. The process of digestion is aided by multiple enzymes which are mainly derived from the stomach, pancreas and small intestine. A delicate flow of either hydrogen or bicarbonate ions allows the enzymes to function under optimal conditions, and bile acids aid in the digestion and absorption of fat. This complex series of events is influenced not only by the presence of food in the gastrointestinal tract, but also by nervous and humoral influences. Both parasympathetic, mainly the vagus, and sympathetic nerve supplies to the gut are important and more central mechanisms such as the hypothalamus exert notable influences. The gut harbours many hormones: amines, prostaglandins, and a complex group of closely interrelated polypeptide hormones. The gut cannot be considered as an isolated organ in the body as there is increasing evidence that the gut can influence many distant organs and cells, not only by the absorption of nutrients but also by its numerous hormones. The gut can be considered as an endocrine organ just as the pituitary or the adrenal, transmitting messages to many areas of the body.

Diabetes mellitus may be regarded as a disturbance in the normal metabolism of ingested foodstuffs predominantly due to a lack of insulin or a lack of effectiveness of insulin. The gut may play a part in this disturbance by several means. The influence of the gut over insulin secretion (the entero-insular axis) by humoral or nervous mechanisms may be altered, resulting in inappropriate amounts of insulin being released. The gastrointestinal tract may suffer damage due to the complications of diabetes mellitus and this would result in malfunction of the gut and a worsening of the diabetic state. Finally oral diabetic therapy may exert much of its influence on gut mechanisms. The role of the gut in diabetes mellitus is shown in Table 3.1.

INSULIN SECRETION IN DIABETES MELLITUS

Before discussing gastrointestinal factors which may regulate insulin secretion, it would be pertinent to discuss briefly the status of insulin secretion in diabetes mellitus. There is general agreement that patients with advanced clinical diabetes have reduced insulin secretion to an oral glucose load[1]. This is especially obvious when the influence of the basal level, which is mainly determined by body weight, is taken into account and the insulin response expressed as relative increments above the basal level[2]. However, there is controversy surrounding the insulin secretion in prediabetes or in the phase before the patient becomes chemically diabetic. Cerasi and Luft, using an intravenous infusion of glucose, have demonstrated that there is an adequate insulin response despite normal glucose tolerance in the healthy monozygotic twin sibs of diabetics[3] and 15-20% of healthy adults[4] and children[5]. Jackson, Mieghem and Keller[6] studied potentially diabetic groups in South Africa, and found that the more potentially diabetic a

Table 3.1 The Gastrointestinal Tract (GI) and Diabetes Mellitus (DM)

1 The effect of DM on the GI tract:
   (a) neuropathy
   (b) autoimmunity
   (c) oral diabetic therapy

2 The effect of the GI tract on DM:
   (a) GI hormones
   (b) nervous mechanisms

group became, the higher was the fasting insulin and the insulin response to oral glucose. However, when chemical diabetes appeared this was associated with a reduced and delayed insulin response. They concluded that the earliest biochemical lesion is associated with insulin excess rather than insulin deficiency. Support for this concept that insulin excesss is the initial lesion in diabetes mellitus comes from Reaven and co-workers[7][8]. Reaven, Olefsky and Farquhar[9] also showed that in patients with chemical diabetes there were hyperinsulinaemic but normal glucose responses to a formula diet which resembled the constituents of a normal diet. Cerasi, Efendic and Luft[10] have responded to the challenge that their intravenous stimuli are unphysiological by employing oral dose-responses to glucose. They concluded that the insulin response to orally administered glucose was inadequate in prediabetic and diabetic subjects.

It is difficult to reconcile all these conflicting data. Certainly the subjects studied are often widely different. Jackson, Mieghem and Keller[6] studied a Tamil Indian 'family' with a high prevalence of diabetes, and a Cape coloured population where hyperglycaemia was common. For their oral glucose tolerance studies, Cerasi, Efendic and Luft[10] chose as some of the prediabetics nine healthy subjects who had a low insulin response to intravenous glucose, and by these criteria defined them as prediabetic. One may conclude that the insulin response to glucose is deficient in established diabetes, but that the insulin secretory status prior to the individual developing diabetes remains largely unknown. The evidence suggesting hypoinsulinism in prediabetes is balanced by numerous papers suggesting hyperinsulinism. The conflict is most probably a result of the ill-definition of prediabetes and consequently ill-defined groups of subjects studied by different types of test.

GASTROINTESTINAL FACTORS IN THE INSULIN RESPONSE

Soon after the discovery by Bayliss and Starling[11] of secretin in 1902, physicians in Liverpool Royal Infirmary were employing secretin-containing extracts of hog duodenal mucosa in the therapy of patients with diabetes mellitus in the hope that it might enhance endogenous insulin production[12]. Dupre[13] revived interest in this field when in 1964 he showed that an intestinal extract significantly increased the rate of disappearance of an intravenously administered glucose load. In the same year McIntyre, Holdsworth and Turner[14] showed that higher plasma insulin levels were obtained following an intrajejunal load of glucose than an equivalent intravenous one. These results have since then been adequately confirmed by Perley and Kipnis[15]. Portocaval anastomosis[14] did not abolish the enhancement of insulin release following an oral glucose load, so that it appeared that intestinal factors rather than hepatic ones were responsible for the effect. Similar studies with respect to amino acids have also shown that intestinal factors exert a greater effect on insulin secretion following oral amino acids as compared with intravenous administration[16].

The gastrointestinal tract may control insulin secretion either by nervous influences or by secretion of the numerous hormones present in the gut. The nervous mechansims will be examined first.

Nervous influences
There is evidence that the parasympathetic nervous system regulates insulin output. Acetylcholine or its analogues have been found to increase insulin release both *in vitro* and *in vivo*[17][18]. Stimulation of the vagus nerve has been shown to release insulin[19]. Vagotomy decreases the insulin release to oral and intravenous glucose in rats[21]. That these are probably direct effects is supported by the use of electron microscopy and special stains which show parasympathetic and sympathetic nerve endings close to the hormone-producing cells of the endocrine pancreas[22-24]. The sympathetic nervous system is also involved as norepinephrine, the sympathetic neurotransmitter, has been found to inhibit insulin release by stimulation of the alpha adrenergic receptor[17][25]. The experiments of Porte *et al.*[26] in which the effect on insulin secretion, of direct stimulation of the mixed autonomic nerve to the pancreas was studied strongly suggest that the sympathetic nerve fibres inhibited insulin secretion and the parasympathetic fibres stimulated insulin secretion. Hommel, Fischer and Retzlaff *et al*[27] showed that sham feeding of a glucose meal in conscious trained dogs bearing doubled-barrelled fistulae of the oesophagus or the stomach resulted in no rise in blood sugar but in considerable increases in their insulin levels. These experiments would suggest that a cephalic phase of insulin secretion exists. Glucagon also appears to be under nervous control. Stimulation of the sympathetic nerves in calves results in glucagon release[28].

The gastrointestinal hormones
Table 3.2 gives an incomplete list of the polypeptide gastrointestinal hormones. The list is incomplete because many other materials have not yet achieved the status of a hormone. Indeed some of the 'hormones' in the list can scarcely be classified as hormones but will be discussed as special studies have been made of them in relation to insulin secretion.

Table 3.2 Gastrointestinal hormones

Gastrin
Cholecystokinin-pancreozymin
Secretin
Glucagon (pancreatic)
Glucagon-like-immunoreactivity (gut)
Gastric inhibitory polypeptide
Vasoactive intestinal polypeptide
Motolin
Insulin-releasing polypeptide

Considerable structural homology exists amongst these hormones. Gastrin and cholecystokinin-pancreozymin (Figure 3.1) share the same biologically active site in the C-terminal tetrapeptide amide Try-Met-Asp-Phe-NH_2. Secretin, glucagon, vasoactive intestinal polypeptide (VIP) and gastric inhibitory polypeptide (GIP) (Figure 3.2) also show considerable structural overlap.

Gastrin

Gastrin which has been isolated from the antral mucosa of man, hog, dog and sheep[29] is a polypeptide of 17 amino acid residues (Figure 3.1). In all species gastrin occurs in two forms, gastrin I without, and gastrin II with ethereal sulphate on the ring of tyrosine in position 12. Although the two gastrins have different physico-chemical properties, they apparently do not differ in biological activity. The major action of gastrin is to stimulate gastric acid secretion[30].

A radioimmunoassay for gastrin was first reported by McGuigan[31] in 1968 and since then by Yalow and Berson[32] in 1970. Cross reaction is encountered with cholecystokinin-pancreozymin (CCK-PZ) but this is so slight that it can be discounted in plasma assays.

Cholecystokinin-pancreozymin (CCK-PZ)

Two separate hormones were postulated on the basis of physiological studies but chemical purification by Mutt and Jorpes[33] in 1967 has shown that all the actions reside in a single molecule. CCK-PZ is a polypeptide of 33 amino acid units[34] and is located in the upper small intestine. Interest is mainly focused on the C-terminal end of CCK-PZ which, as discussed above, is similar to gastrin and this similarity in structure is probably

Figure 3.1. Amino Acid sequence of porcine cholecystokinin-pancreozymin (CCK-PZ) and gastrin II (G II)

CCK-PZ

1 2 3 4 5 6 7 8 9 10 11 12 13 14 15 16

Lys-Ala-Pro-Ser-Gly-Arg-Val-Ser-Met-Ile-Lys⁺-Asn-Leu-Gln-Ser-Leu-

17 18 19 20 21 22 23 24 25 26 27 28 29 30 31 32 33

Asp-Pro-Ser-His-Arg⁺-Ile-Ser-Asp-Arg⁺-Asp-Tyr(SO₃)-Met-Gly-Trp-Met-Asp-Phe(NH₂)

G II

1 2 3 4 5 6 7 8 9 10 11 12 13 14 15 16 17

Glu-Gly-Pro-Tyr-Met-Glu-Glu-Glu-Glu-Ala-Tyr(SO₃)-Gly-Trp-Met-Asp-Phe(NH₂)

Figure 3.2 Amino acid sequence of porcine vasoactive intestinal peptide (VIP), secretin (S), glucagon (G), and gastric inhibitory peptide (GIP)

	1	2	3	4	5	6	7	8	9	10	11	12	13	14	15	16	17	18	19	20	21
VIP	His-	Ser-	Asp-	Ala-	Val-	Phe-	Thr-	Asp-	Asn-	Tyr-	Thr-	Arg-	Leu-	Arg-	Lys-	Gln-	Met-	Ala-	Val-	Lys-	Lys-
S	His-	Ser-	Asp-	Gly-	Thr-	Phe-	Thr-	Ser-	Glu-	Leu-	Ser-	Arg-	Leu-	Arg-	Asp-	Ser-	Ala-	Arg-	Leu-	Gln-	Arg-
G	His-	Ser-	Gln-	Gly-	Thr-	Phe-	Thr-	Ser-	Asp-	Tyr-	Ser-	Lys-	Tyr-	Leu-	Asp-	Ser-	Arg-	Arg-	Ala-	Gln-	Asp-
GIP	Tyr-	Ala-	Glu-	Gly-	Thr-	Phe-	Ile-	Ser-	Asp-	Tyr-	Ser-	Ile-	Ala-	Met-	Asp-	Lys-	Ile-	Arg-	Gln-	Gln-	Asp-

	22	23	24	25	26	27	28	29	30	31	32	33	34	35	36	37	38	39	40	41	42	43
VIP	Tyr-	Leu-	Asn-	Ser-	Ile-	Leu-	Asn(NH_2)															
S	Leu-	Leu-	Gln-	Gly-	Leu-	Val(NH_2)																
G	Phe-	Val-	Gln-	Trp-	Leu-	Met-	Asn-	Thr														
GIP	Phe-	Val-	Asn-	Trp-	Leu-	Leu-	Ala-	Gln-	Gln-	Lys-	Gly-	Lys-	Lys-	Ser-	Asp-	Trp-	Lys-	His-	Asn-	Ile-	Thr-	Gln

responsible for an overlap of physiological functions. The principle actions of CCK-PZ are to contract the gall bladder[35] and to stimulate the secretion of enzymes in pancreatic juice[36]. However it has been demonstrated that CCK-PZ and gastrin share certain physiological properties, in that both hormones stimulate gastric acid secretion and pancreatic enzyme secretion[37-40]. Grossman[41] suggested that CCK-PZ and gastrin act on one receptor.

Pancreatic glucagon

In 1924 Kimball and Murlin[42] noted that when most commercial insulin preparations were injected intravenously, the typical hypoglycaemic effects were preceded by a short period of hyperglycaemia. This fleeting rise in glucose concentration was due to the presence of an impurity, named 'glucagon' or 'mobiliser of sugar' by Collens and Murlin[43] in 1929, who first investigated its properties and attempted its purification. Staub, Sinn and Behrens[44] in 1955 eventually purified and crystallised glucagon. Glucagon has a molecular weight of 3485 and is a straight chain polypeptide containing 29 amino acid residues. The hormone was synthesised by Wunsch[45] in 1967. Fluorescent antibody studies have demonstrated that glucagon is secreted from the α cells of the islets of Langerhans[46].

Glucagon has glycogenolytic[47], gluconeogenic[48], and lipolytic[49] properties. Perhaps the most exciting and even surprising new property for glucagon has been its ability to stimulate insulin secretion independently of hyperglycaemia produced[50 51]. Many of the actions of glucagon are mediated by activation of adenylcyclase on cell membranes and the resultant generation of the cyclic nucleotide, cyclic 3′,5′-AMP, the second messenger[52], which in turn produces a physiological response.

Glucagon-like immunoreactivity (GLI) from gut

The radioimmunoassay for pancreatic glucagon resulted in the discovery of GLI in the intestine. It was noted that oral glucose increased immunoreactive glucagon (IRG) levels in the blood[53]. Further studies suggested that this IRG was of intestinal rather than pancreatic origin[54]. Earlier studies[55] had indicated that extracts of intestine contained material with the biological activities of glucagon. The radioimmunoassay for glucagon has amply confirmed that there is IRG in the intestine[56], large amounts being found in jejunum, ileum and colon, with smaller amounts in the stomach and duodenum. Indeed the human rectum[57] contains rather large amounts of the material. The origin of the blood levels of IRG after oral glucose was solved by Buchanan *et al.*[57] who showed that intrajejunal glucose in pancreatectomised dogs gave rise to striking increases of IRG in the blood. In an acute experiment in a single dog, Buchanan[58] showed that pancreatectomy resulted in a reduction in circulating IRG levels, and when the stomach and small bowel were also removed the plasma IRG virtually reached zero. Removal of the colon had no further effect. Because of obvious confusion with pancreatic glucagon, Unger has coined the descriptive expression 'glucagon-like immunoreactivity' (GLI) to stress that the material, perhaps only by chance, cross-reacts in a glucagon radioimmunoassay, and that there the resemblance to pancreatic glucagon may end.

However, because of the inevitable link-up with pancreatic glucagon, attention during purification and characterisation processes has been directed towards differences from or similarities to pancreatic glucagon. Unger and colleagues, in two recent publications[59][60] have found on gel filtration that GLI resides in two peaks, peak II having a molecular weight similar to pancreatic glucagon and peak I being considerably larger. They found glycogenolytic activity to reside in the peak II molecule but not in peak I and they found all the products to show different immunological characteristics when studied with a specific antiserum to pancreatic glucagon. Murphy, Buchanan and Elmore[61], using affinity chromatography (solid phase glucagon antibodies) produced highly purified fractions of GLI from pork ileum. On gel filtration they found 90% to reside in a peak of molecular weight 12 000 (large GLI) and 10% to reside in a peak of molecular weight 3500 (small GLI). Further studies on large GLI showed that it resolved into three fractions on polyacrylamide gel electrophoresis and three N-terminal amino acids. Biological testing of large GLI proved negative. Antibodies to pancreatic glucagon can be produced which show discrimination between pancreatic glucagon and gut GLI[62] in radioimmunoassays. Such antibodies have been further defined in that antibodies which discriminate between pancreatic glucagon and gut GLI are orientated towards the C-terminus of pancreatic glucagon[63]. When purified fractions of gut GLI are reacted with these antibodies, large GLI reacts only with N-terminal reactive antibodies but small GLI shows reaction with both antibodies[64]. This would suggest that gut GLI exists in several molecular forms and, immunologically, small GLI most resembles pancreatic glucagon. Large GLI does not appear to resemble pancreatic glucagon with respect to biological activity and it may be that the biological activity resides in small GLI.

Secretin has been isolated from pig jejunum by Jorpes*et al.*[65] in 1962. Hog secretin has 27 amino acid residues. Like gastrin, its C-terminal residue is an amide. No analogues or fragments of the molecule with secretin-like activity have been found. The structure of secretin bears a striking similarity to glucagon. No fewer than 14 of the 27 amino acid units of secretin occupy the same position as in glucagon, counting from the N-terminal end. However, despite this similarity, antibodies to glucagon do not react with secretin[66]. Secretin stimulates an increase in volume flow and electrolyte content of the pancreatic juice without stimulating the release of digestive enzymes[67]. Wormsley and Grossman[68] have shown that purified secretin is an inhibitor of gastrin-stimulated gastric secretion in dogs and Jorpes *et al.*[69] showed that secretin stimulated bile secretion.

Gastric inhibitory polypeptide

A pure polypeptide chemically distinct from any known hormone has been isolated from pig duodenum by Brown, Mutt and Pederson[70] in 1970. Its structure was determined by Brown and Dryburgh[71] in 1971, and its physiological properties were claimed to correspond with those of enterogastrone[72], i.e. inhibition of gastric acid secretion and motility following fat and possibly by acid and hypertonic solutions. The hormone is referred to as gastric inhibitory polypeptide (GIP).

Vasoactive intestinal polypeptide

This polypeptide has been isolated from the porcine upper intestinal wall by Said and Mutt[73] in 1972. In addition to exhibiting several biological activities which may be attributed to a relaxant effect on vascular and systemic smooth muscle, it stimulates exocrine pancreatic function in a secretin-like fashion, with an apparent efficiency on a molar basis, of 5–10 % of that of secretin and also gives rise to hyperglycaemia with about 30% of the efficiency of glucagon[73].

Motolin

Motolin has been purified and the amino acid sequence elucidated. It is a straight chain polypeptide of 22 amino acid residues[74]. The physiological actions of motolin are to stimulate motor activity in the body and antrum of the stomach and in the duodenal bulb.

THE ROLE OF GASTROINTESTINAL HORMONES IN THE INSULIN RESPONSE

At least four criteria need to be satisfied before concluding that a gut hormone has a physiological role in the control of insulin secretion:

(1)   the hormone must be shown to stimulate insulin release;
(2)   stimulation of insulin release should occur at physiological levels
      of the gut hormone;
(3)   the gut hormone must be pure;
(4)   the gut hormone should show a rise in the bloodstream during the
      physiological event under consideration.

Each gut hormone will be discussed in the light of these various criteria
in the succeeding paragraphs.

Secretin

Several studies have revealed that secretin can stimulate insulin release.
Unger *et al.*[75] in 1966 showed that secretin gave a threefold rise in mean
pancreaticoduodenal vein plasma insulin concentration, and that this rise,
unlike that of glucagon, was unaccompanied by arterial hyperglycaemia.
These studies were extended to humans by Jarrett and Cohen[76] who
showed a sharp and rapid insulin response to infusion of secretin. That
this effect of secretin might be pharmacological was suggested by the
experiments of Buchanan *et al.*[77] who could show no effect on insulin
secretion in anaesthetised dogs when secretin was given in doses sufficient to
stimulate a copious flow of pancreatic juice. The effect of secretin is only
noted in the presence of the exocrine pancreas[78] as secretin does not
stimulate insulin secretion in isolated islets of Langerhans[66]. That the
insulinotropic action of secretin might have physiological significance has
however been suggested by Dupre *et al.*[16] who reported that intestinal
acidification did result in release of insulin. The establishment of a
radioimmunoassay of secretin by Young *et al.*[79] has considerably facilitated
further studies in this area. These workers have shown[80][81] that serum
secretin levels rose rapidly after oral glucose or protein and preceded the
elevation of serum insulin. They were also able to show that intravenous
infusion of secretin caused a release of insulin when serum secretin levels
were within the physiological range. However, Buchanan *et al.*[82], also using
a radioimmunoassay for secretin have shown that plasma secretin levels fall
after oral or intraduodenal glucose in man and dogs.

   Different radioimmunoassay techniques may account for the
discrepancy. Chisholm *et al.*[80] raised antibodies to impure antigen which
they also used as standard, although their labelled antigen was pure.
Buchanan on the other hand[83] used purified materials throughout. The
integrity of the Chisholm *et al.*[80] assay depends greatly on the absolute
purity of the labelled hormone, as any contamination could result in their
impure standards or antibodies giving artefactual results. It is also possible

that antibodies of differing specificities may give totally different results, as has been seen in the glucagon story with cross-reacting and pancreas-specific antisera. Previous workers using bioassays suggest that glucose does not stimulate secretin release[84] [85].

Secretin cannot be considered to be a strong candidate for a role in the entero-insular axis for glucose. It is dubious if physiological concentrations stimulate insulin release and the hormone appears to fall after glucose rather than to rise. However a role for secretin in the insulin rise after protein cannot be ruled out.

Cholecystokinin-pancreozymin (CCK-PZ)

Dupre and Beck[86] showed that an extract of intestinal mucosa, biologically indistinguishable from CCK-PZ, enhanced the increase in blood-insulin concentration which occurred in response to the intravenous infusion of glucose. CCK-PZ can however also stimulate the release of pancreatic glucagon in dogs[77] [87]. Füssganger et al.[88], utilising the isolated perfused rat pancreas, suggested that insulin release was secondary to the glucagon release. Using a radioimmunoassay for CCK-PZ, Young, Lazarus and Chisholm[89] have shown a rise of plasma CCK-PZ following oral glucose, although the rise was delayed as compared with insulin and secretin. CCK-PZ is regarded as a protein-responsive hormone, protein products and amino acids being the most potent known stimuli to its release[85]. Ohneda et al.[90] therefore studied the effect of CCK-PZ on an amino acid infusion and found that it was capable of augmenting the islet-cell responses to hyperaminoacidaemia. Amino acids increase insulin release without hypoglycaemia[91] and it may therefore be that glucagon is secreted after a predominantly protein meal to prevent the hypoglycaemia which might otherwise result from the protein-stimulated insulin secretion.

However, the majority of these experiments were conducted using 10% pure CCK-PZ, a preparation which is now known to be rich in gastric inhibitory polypeptide. Until data are forthcoming utilising pure CCK-PZ it cannot be said whether this hormone has any part to play in the entero-insular axis.

Gastrin

Gastrin has slight and transient effects on insulin secretion in dogs and no effect on glucagon secretion[92]. It appears unlikely therefore that gastrin is seriously implicated in stimulating the release of islet hormones. In addition, in various human studies, no correlation could be found between the levels of gastrin and insulin[93]. Plasma gastrin levels show only slight and transient elevation after oral glucose in man[94] but gastrin secretion is actually suppressed by glucose using an *in vitro* model of rat antral tissue[95].

Glucagon and gut glucagon-like immunoreactivity (GLI)

Pancreatic glucagon stimulates insulin secretion both *in vivo*[50 51] and *in vitro*[96 97]. That these findings might have physiological significance has been suggested by the results of Ketterer, Eisentraut and Unger[98] and Buchanan *et al*.[99], who showed that tiny 'physiological' doses of glucagon administered via the portal vein in dogs could increase insulin secretion. As investigators primarily with an interest in the endocrine secretions of the pancreas have been studying gut GLI, it is scarcely surprising that this material has come under examination as a possible mediator of the intestinal effect upon insulin secretion. The results in this direction are conflicting at the present time, mainly because of differences in the purity of GLI and different 'insulin-secreting models' used to study its effects. Murphy *et al*.[67] who produced a highly purified gut GLI could not demonstrate that the large molecular form had any insulin-stimulating activity.

Gastric inhibitory polypeptide (GIP)

The following data concerning GIP have been obtained from Brown[100]. GIP which is uncontaminated with other known gastrointestinal hormones has been found to be a powerful stimulator of insulin secretion in man and dog and this release is associated with a fall in blood sugar, and indeed GIP produces a greater insulin response than glucose in the dog. GIP also rises after oral glucose. These data must place GIP in a position of importance in the entero-insular axis for glucose. In addition physiological concentrations of GIP have now been shown to stimulate insulin release in man[101].

Vasoactive intestinal peptide (VIP), motolin and insulin-releasing polypeptide (IRP)

Motolin does not appear to release insulin[100] and VIP also appears to be poorly active in this respect[102].

Turner[103] has suggested the existence of another polypeptide which may regulate insulin secretion after glucose. He has called this material 'insulin-releasing polypeptide' (IRP). Moody *et al*.[104] also discuss the existence of insulin-releasing activity (IRA) of the intestine separately from known intestinal hormones. Turner and Marks[105] have shown that partially purified IRP enhanced glucose-induced insulin release in the rat *in vivo*. These studies and claims were made before the existence of GIP and its potent insulin-releasing properties were known, and IRP and IRA must be separated clearly from GIP before they can be taken seriously as insulin-releasing hormones.

GASTROINTESTINAL HORMONES IN DIABETES MELLITUS

This has been studied in two ways. Firstly, the effect of gastrointestinal hormones on insulin secretion has been studied in diabetics, and secondly the circulating levels of gastrointestinal hormones have been measured in diabetic patients by radioimmunoassay.

Several papers have reported that secretin-induced insulin release in diabetic patients may be preserved and yet such patients may show markedly reduced insulin secretion to glucose or tolbutamide[106-111].

In patients with exocrine pancreatic insufficiency, yet with acceptable endocrine pancreatic function, striking reduction in the insulin release following secretin is noted[112,113]. This effect has also been induced experimentally in rabbits by preserving endocrine function but destroying exocrine function by ligating the duct of Wirsung[114,115]. Deckert *et al.*[112] speculate that secretin-induced shift in the pH of the pancreatic tissue secondary to the bicarbonate secretion might be responsible for the secretion of insulin, either by an inhibition of the sodium pump of the β-cells or promoting an influx of calcium ions into the β-cells. The reduced insulin secretion following secretin in patients with reduced exocrine pancreatic function is a consequence of the low exocrine secretion of bicarbonate in these patients. Pfeiffer *et al.*[116] consider that an 'entero-receptor' of the β-cell is stimulated via special intrapancreatic exocrine-endocrine channels by some chemical messenger released from the primary target residing in the exocrine pancreatic tissue. Lerner and Porte[117] suggest that glucose and secretin stimulate functionally separate storage pools of insulin but that the acute response to either stimulus is partly determined by exposure to the other.

Few, if any, studies have appeared on secretin, gastrin and other gut hormone levels in diabetes mellitus, although in recent years glucagon secretion in diabetes mellitus has been studied. Most publications report an abnormality of circulating glucagon in diabetes mellitus. Several publications have appeared from Unger's laboratory. They have tended to show higher glucagon secretion after arginine in diabetes and a lack of suppressibility of glucagon in diabetes to glucose[118,119]. Wise, Hendler and Felig[120] have shown higher basal glucagons in diabetics who are not overweight and an increased glucagon response to intravenous alanine. Buchanan and McCarroll[121] have shown that some diabetics show a rise in glucagon secretion to glucose (instead of a normal fall) and that this phenomenon is associated with insulin deficiency. Buchanan and Mawhinney[122] [123] have shown that glucagon release is increased from insulin-deficient rat islets and that this increase is corrected by the addition of insulin. Trimble, Hadden and Montgomery[124] have found that glucagon levels are raised mainly in thin diabetics and early results in a small number of diabetics would suggest that dietary management of their diabetes results in correction of these glucagon abnormalities. Ardill, Montgomery and Hadden[125] have found plasma gastrin levels to be elevated in some diabetics and the gastrin response of the diabetics to oral glucose to correlate inversely with insulin secretion.

The source and nature of circulating glucagon levels as measured by radioimmunoassay have been a source of difficulty over the years. When a radioimmunoassay for glucagon was first described it was soon found that the assay measured not only glucagon in the pancreas but also a glucagon-like immunoreactive (GLI) substance in the gut[54] [56]. Antibodies to pancreatic glucagon can be produced which show discrimination between pancreatic glucagon and gut GLI[62]. It is therefore considered possible to measure pancreatic glucagon specifically. However, recent studies may suggest that the concept of a specific pancreatic glucagon assay is erroneous. Completely pancreatectomised dogs show high levels of glucagon as measured by a so-called specific antiserum[126]. It is likely that the relationships between pancreatic glucagon and gut GLI are extremely complex. Highly purified gut GLI resolves itself into several fractions[61] and one of these fractions reacts with 'pancreas-specific' antibodies[63]. In addition there is some evidence for conversion of a gut GLI fraction into a pancreatic glucagon-like material[64]. This leaves the interpretation of the glucagon levels in diabetes open to the question that the glucagon may originate not from the pancreas but from the gut.

TUMOURS SECRETING PANCREATIC AND GUT HORMONES IN DIABETES MELLITUS

Insulinoma

It appears almost contradictory that patients with insulin-secreting tumours may demonstrate a diabetic response in an oral glucose tolerance test[127]. This may be attributed to autonomous secretion of insulin or to an abnormal insulin or to the 'Somogyi' effect[128]. However it is unlikely that the situation would give rise to great difficulty in differentiating the condition from diabetes mellitus.

Verner Morrison Syndrome

This is sometimes referred to as the watery diarrhoea hypokalaemic achlorhydric syndrome (WDHA) and was first described by Verner and Morrison[129] in 1957. Nine out of 17 patients had hyperglycaemia[130] but this derangement in carbohydrate tolerance may be the known effect of severe hypokalaemia in causing alteration in glucose metabolism[131][132]. It has recently been suggested that these tumours may secrete either gastric inhibitory polypeptide (GIP)[133] or vasoactive intestinal peptide (VIP)[134]. The evidence appears stronger however that the tumours secrete VIP rather than GIP as the specificity of the GIP antiserum used in the Elias *et al.*[133] report was questionable and the positive result may have been due to contaminating VIP antibodies[134]. If VIP is secreted by these tumours then this hormone may be responsible for the mild diabetic syndrome as it has been shown to produce glycogenolysis and hyperglycaemia[135].

Glucagonomas

McGavran *et al.*[136] reported the first tumour proven by radioimmunoassay to be secreting glucagon. The patient presented with mild diabetes mellitus and a bullous eczematoid dermatitis, and an α-cell carcinoma of the pancreas was present at autopsy. Church and Crane[137] described a similar patient with autopsy findings of a pancreatic islet cell tumour of α-cell type and referred to an earlier patient described by Becker, Kahn and Rothmann[138] in 1942. Yoshinaga *et al.*[139] described a patient with an α-cell adenoma and diabetes mellitus. Mallinson *et al*[140] describe an extensive series of nine patients with glucagonomas. These patients had a necrolytic migratory erythema, stomatitis, weight loss and in seven cases, diabetes. All nine were found to have pancreatic tumours. In all cases the diagnosis was clinched by radio-immunoassay of glucagon in the blood or tumour. In the one case in which the tumour was resected the patient recovered completely. Gleeson *et al.*[141] described a patient with marked gastrointestinal abnormalities and renal tumour which was found to contain cells histologically similar to the pancreatic α-cells. This tumour was rich in glucagon-like immuno-reactivity (GLI) and it appeared from immunological studies and molecular weight to resemble more the GLI of gut rather than pancreas[142]. This patient also had mild diabetes mellitus.

Other gut endocrine tumours

Apart from the well known Zollinger-Ellison tumours which produce gastrin, tumours producing secretin, CCK-PZ, GIP or motolin have not been described. Zollinger-Ellison patients do not have diabetes mellitus.

THE EFFECT OF DIABETES MELLITUS ON THE GASTRO-INTESTINAL TRACT

The stomach

There is an increased incidence of pernicious anaemia and achlorhydria in diabetes mellitus. Achlorhydria was present in 39% of the patients studied by Rabinowitch[143]. That this might be due to an association between diabetes and diseases characterised by organ-specific autoimmunity was suggested by the studies of Irvine *et al.*[144] These authors showed a significantly increased incidence of gastric parietal cell antibodies in the sera of diabetics compared with controls. It would be expected from these results that there would be an increased incidence of pernicious anaemia in diabetes mellitus, and the literature would tend to support this. The incidence appears to be about ten times that of the normal population[145−147]. The incidence however might be greater if account were taken of the higher incidence of pernicious anaemia in the elderly. Irvine *et al.*[144] found an increased incidence of intrinsic factor antibody in the sera of female diabetics over 40 years and of those who were further studied, evidence of latent pernicious anaemia was found in the majority.

The incidence of peptic ulcer in diabetics is lower than in the normal population[145 148]. That this may be due to the lowered gastric acid secretion in diabetes must be considered a possibility.

Impaired motility of the stomach (gastro-paresis) occurs in diabetes mellitus[149 150]. It is likely that this is due to autonomic neuropathy as one third of these patients also have autonomic diabetic diarrhoea[151]. Episodes of intractable vomiting occur in some diabetic patients and this appears to be a separate entity from gastro-paresis. A barium meal in such patients passes easily through the pylorus but filling of the first part of the duodenum may be very difficult and the small intestine is not outlined[151].

The gut

Diabetic diarrhoea is usually seen in severe diabetics requiring insulin and who have definite evidence of autonomic neuropathy[151]. The diarrhoea is frequently nocturnal. Faecal fat excretion and jejunal biopsy are usually normal[152 153]. A barium meal frequently shows delay in gastric emptying[154] and transit through the small intestine is usually normal or rapid[151]. The diarrhoea is probably due to autonomic neuropathy but abnormal bacterial growth in the gut may be another factor[154]. A speculative suggestion is that a disturbance in gastrointestinal hormones in diabetes mellitus (see above) may result in motility abnormalities.

The exocrine pancreas
Some patients with diabetes mellitus may have an underlying pancreatic disease such as pancreatitis or carcinoma which has caused the diabetic state. Less certain however is the functional state of the exocrine pancreas in patients with idiopathic diabetes mellitus. The studies of Baron and Nabarro[155] would suggest that an abnormality of pancreatic exocrine function is frequently found, but this abnormality gives rise to no symptoms and does not develop over the years. Such a high incidence of exocrine abnormality in diabetes would be challenged by other workers who found normal function[156-158].

THE EFFECT OF INSULIN AND DIABETIC DRUGS ON THE GUT
Insulin
The effect of insulin on the intestine is controversial. In states of insulin deficiency such as produced with alloxan[159], and pancreatectomy[160], glucose transport in the intestine is enhanced. Apparently the increased absorption of glucose could be reversed by administration of insulin[159]. However, Dubois and Roy[161] found a significant increase in sugar absorption when insulin was administered intra-arterially to subjects treated with anti-insulin. Love and Canavan[162] found that insulin increased glucose absorption in *in vitro* rat-intestinal-sac preparations.

Sulphonylureas
Teale and Love[163] found that tolbutamide and glibenclamide reduced active glucose transfer by sacs of everted rat jejunum. They considered that sulphonylureas disrupted intracellular supplies of ATP, thereby affecting the intestinal transport mechanism. However Bolros *et al.*[164] found in intact rats that sulphonylureas increased galactose transport in the gut, but that these results may have been due to the action of the drugs on insulin secretion. In addition studies in which attempts were made in man to detect impairment of gut glucose transport by glibenclamide were foiled by the significant effect of the drug on insulin secretion[166]. It cannot be answered whether sulphonylureas alter gut glucose transport in patients chronically on these drugs.

Diguanides

There is more uniform opinion concerning the action of these drugs on glucose transport in the gut. In 58 diabetic patients administration of diguanides suppressed intestinal glucose absorption resulting in marked flattening of the glucose response to a standard oral glucose tolerance test[167]. Phenformin pre-treatment flattened the glucose and insulin response to oral glucose but not to intravenous glucose and led to the conclusion that phenformin significantly inhibited intestinal glucose transport[168]. Phenformin may also impair exocrine pancreatic function and the actions of secretin[169]. Creutzfeldt[170] suggests that biguanides inhibit insulin release after oral food ingestion, not only by impairment of absorption but by inhibition of intestinal hormone release.

CONCLUSIONS

The discovery of insulin led to a normal life for many diabetics but did not eventually prevent the remorseless advance of the chronic complications of the disease. The effect of diabetic complications on the gastrointestinal tract has been discussed in this chapter; so too have the gut effects of diabetic drugs and insulin, but it is not by these mechanisms that the disease will be explained.

The advent of radioimmunoassay has allowed a careful examination of insulin secretion in diabetes mellitus. However extensive studies in this area have been confusing, particularly with respect to insulin secretion preceding the development of the disease. Studies in depth are required of the insulin secretory status in a diabetic's life, hopefully during the time before he develops the disease. It is of course difficult to define the area of 'prediabetes' and the multiplicity of different reports of insulin secretion in prediabetes suggests that the authors are studying heterogeneous groups. It is clear however that diabetics eventually develop insulin exhaustion and this is clinically manifested by the syndrome of diabetes mellitus.

Could the gastrointestinal tract exert sufficient control over insulin secretion to be responsible for the eventual breakdown of the insulin secretory mechanism? One must examine both the nervous influences which control pancreatic and gut hormone secretion and the gut hormones themselves. Nervous influences are vitally important in insulin and glucagon secretion, and indeed one group of workers suggests that the entero-insular axis is over-active in diabetes mellitus and this overactivity may be mediated via the vagus[171]. The activity of nervous influences on insulin secretion in diabetes mellitus remains to be tested.

The rapid upsurge of the gastrointestinal tract as a major endocrine organ has owed much to the gigantic efforts of Jorpes and Mutt. Hormones such as gastrin and motolin which have little effect on insulin secretion and carbohydrate metabolism can be discounted as having a role in the aetiology of diabetes. Also, one must for the present ignore poorly characterised materials such as gut GLI and insulin-releasing peptides. From the point of view of the pathogenesis of diabetes mellitus, attention must be focused on fully characterised hormones which have important effects on insulin secretion and carbohydrate metabolism. These include gastric inhibitory polypeptide (GIP), secretin, pancreatic glucagon, and vasoactive intestinal peptide (VIP). A cloud for the moment hangs over the role of cholecystokinin-pancreozymin (CCK-PZ) in insulin secretion as so much data has been confused by the presence in preparations of CCK-PZ of the potent insulin releaser GIP. At the present time GIP fulfils many of the criteria required to be the major hormone in the control of the entero-insular axis, but so far all the data concerning this hormone emanate from a single laboratory and clearly this important work needs confirmation and expansion. Radioimmunoassays for many of these gut hormones have been developed but frequent disparity in results from laboratory to laboratory suggests difficulties in standardisation and specificity of these assays. Full understanding of the results of such assays may only be achieved once tissue and circulating forms of gut hormones have been characterised and the specificity of the assays further defined. The radioimmunoassays should preferably be backed by sensitive bioassay methods and receptor site assays may have much to offer in this respect. There remains a considerable amount of further development before a definitive statement can be levied that the entero-insular axis has a part to play in the development of diabetes mellitus. However some gut hormones may have a role in certain manifestations of the disease, e.g. the insulin sensitivity of the juvenile diabetic may be due to poor glucagon responses to hypoglycaemia[172].

Sometimes hormone-secreting tumours may help to elucidate the role of the hormone in the pathogenesis of disease. The glucagonoma syndrome is frequently associated with diabetes mellitus but the authors of a recent extensive description of the disease suggest that the very high glucagon levels associated with the mild diabetes would not lend support that minor abnormalities of glucagon may play a part in the pathogenesis of idiopathic diabetes mellitus[140]. However, it does not exclude that minor abnormalities of glucagon secretion over many years may result in diabetes mellitus. VIP tumours also are associated with mild diabetes mellitus but tumours secreting various other gut hormones still await description.

The role of the entero-insular axis in the aetiology of diabetes mellitus must on present evidence remain speculative. An abnormality in this area could eventually lead to the development of the disease but so far a frank abnormality has not been defined. In such a complex disease as diabetes mellitus which affects so many of the bodily systems, it is often impossible to state whether an abnormality is a primary or a secondary event. The rapidly expanding field of gut endocrinology has impinged on diabetes mellitus and future developments may very well lead to better understanding of this enigmatic disease.

References
1. McKiddie, M. T., Buchanan, K. D. and Hunter, I. A. (1969). Plasma insulin studies in two hundred patients with diabetes mellitus. *Quart. J. Med.*, **38**, 445
2. Bagdade, J. D., Bierman, E. L. and Porte, D. (1967). The significance of basal insulin levels in the evaluation of the insulin response to glucose in diabetic and non-diabetic subjects. *J. Clin Invest.*, **46**, 1549
3. Cerasi, E. and Luft, R. (1967). Insulin response to glucose infusion in diabetic and non-diabetic monozygotic twin pairs. Genetic control of insulin response? *Acta Endocrinol., Copenhagen*, **55**, 330
4. Cerasi, E. and Luft, R. (1967). The plasma insulin response to glucose infusion in healthy subjects and in diabetes mellitus. *Acta Endocrinol., Copenhagen*, **55**, 278
5. Cerasi, E. and Luft, R. (1970). The occurrence of low insulin response to glucose infusion in children. *Diabetologia*, **6**, 85
6. Jackson, W. P. U., Van Mieghem, W. and Keller, P. (1972). Insulin excess as the initial lesion in diabetes. *Lancet*, **i**, 1040
7. Reaven, G. M., Silvers, A. and Farquhar, J. W. (1970). Study of relationship between plasma insulin concentration and efficiency of glucose uptake in normal and mildly diabetic subjects. *Diabetes*, **19**, 571
8. Reaven, G. M., Shen, S. W., Silvers, A. and Farquhar, J. W. (1971). Is there a delay in the plasma insulin response of patients with chemical diabetes mellitus? **20**, 416
9. Reaven, G. M., Olefsky, J. and Farquhar, J. W. (1972). Does hyperglycaemia or hyperinsulinaemia characterise the patient with chemical diabetes. *Lancet*, **i**, 1247
10. Cerasi, E., Efendic, S. and Luft, R. (1973). Dose-response relation between plasma-insulin and blood glucose levels during oral glucose loads in prediabetic and diabetic subjects. *Lancet*, **i**, 794
11. Bayliss, W. M. and Starling, E. H. (1902). The mechanism of pancreatic secretion. *J. Physiol. (Lond.)*, **28**, 325
12. Moore, B., Edie, E. S. and Abram, J. H. (1906). On the treatment of diabetes mellitus by acid extract of duodenal mucous membrane. *Biochem. J.*, **1**, 28

13. Dupre, J. (1964). An intestinal hormone affecting glucose disposal in man. *Lancet*, **ii**, 672

14. McIntyre, M., Holdsworth, C. D. and Turner, D. S. (1965). Intestinal factors in the control of insulin secretion. *J. Clin. Endocrinol.*, **25**, 1317

15. Perley, M. J. and Kipnis, D. M. (1967). Plasma insulin response to oral and intravenous glucose: studies in normal and diabetic subjects. *J. Clin. Invest.*, **46**, 1954

16. Dupre, J., Curtis, J. D., Unger, R. H., Waddell, R. W. and Beck, J. C. (1969). Effects of secretin, pancreozymin or gastrin on the response of the endocrine pancreas to administration of glucose or arginine in man. *J. Clin. Invest.*, **48**, 745

17. Malaisse, W., Malaisse Lagae, F., Wright, P. H. and Ashmore, J. (1967). Effects of adrenergic and cholinergic agents upon insulin secretion *in vitro. Endocrinology*, **80**, 975

18. Kajinuma, H., Kaneto, A., Kuzuya, T. and Nakao, K. (1968). Effects of methacholine on insulin secretion in man. *J. Clin. Endocrinol. Metab.*, **28**, 1384

19. Frohman, L. A., Ezdinli, E. Z. and Javid, R. (1967). Effect of vagotomy and vagal stimulation on insulin secretion. *Diabetes*, **16**, 443

20. Kaneto, A., Kosaka, K. and Nakao, K. (1967). Effects of stimulation of the vagus nerve on insulin secretion. *Endocrinology*, **80**, 530

21. Hakanson, R., Liedberg, G. and Lundquist, I. (1971). Effect of vagal denervation on insulin release after oral and intravenous glucose. *Experientia*, **27**, 460

22. Watari, N. (1968). Fine structure of nervous elements in the pancreas of some vertebrates *Z. Zellforsch. Microst. Anat.*, **85**, 291

23. Orci, L., Cameron, D., Lambert, P. E., Kanazawa, Y., Amherdt, M. and Stauffacher, W. (1970). The autonomic nervous system and the β-cell; metabolic and morphological observations made in spiny mice. *Acta Diabetol. Lat.*, **7** *(suppl. 1)*, 184

24. Esterhuizen, A. C., Spriggs, T. L. B. and Lever, J. D. (1968). Nature of islet cell innervation in the cat pancreas. *Diabetes*, **17**, 33

25. Porte, D. (1967). A receptor mechanism for the inhibition of insulin release by epinephrine in man. *J. Clin. Invest.*, **46**, 86

26. Porte, D., Girardier, L., Seydoux, J., Kanazawa, Y. and Posternak, J. (1973). Neural regulation of insulin secretion in the dog. *J. Clin. Invest.*, **52**, 210

27. Homel, H., Fischer, K., Retzlaff, K. and Knöfler, H. (1972). The mechanism of insulin secretion after oral glucose administration. 2. Reflex insulin secretion in conscious dogs bearing fistulas of the digestive tract by sham-feeding of glucose and water. *Diabetologia*, **8**, 111

28. Bloom, S. R. (1973). Glucagon, a stress hormone. *Postgrad. Med. J.*, **49**, 607

29. Grossman, M. I. (1967). Some aspects of gastric secretion. *Gastroenterolgy*, **52**, 882

30. Gregory, R. A. and Tracey, H. J. (1964). The constition and properties of 2 gastrins extracted from hog antral mucosa. *Gut*, **5**, 103

31. McGuigan, J. E. (1968). Immunochemical studies with synthetic human gastrin. *Gastroenterology*, **54**, 1105

32. Yalow, R. S. and Berson, S. A. (1970). Radioimmunoassay of gastrin. *Gastroenterology*, **58**, 1

33. Mutt, V. and Jorpes, J. E. (1967). Contemporary developments in the biochemistry of the gastrointestinal hormone. In: *Recent Progress in Hormone Research*, p.483, (New York: Academic Press)

34. Jorpes, J. E. (1968). The isolation and chemistry of secretin and cholecystokinin. *Gastroenterology*, **55**, 157

35. Fry, A. C. and Oldberg, E. (1928). Hormone mechanism for gall-bladder contraction and evacuation. *Amer. J. Physiol.,* **86,** 599

36. Harper, A. A. and Raper, M. S. (1943). Pancreozymin, a stimulant of the secretion of pancreatic enzymes in extracts of the small intestine. *J. Physiol. (Lond.),* **102,** 115

37. Tracy, H. J. and Gregory. R. A. (1964). Physiological properties of a series of synthetic peptides structurally related to gastrin I. *Nature (Lond.),* **204,** 935

38. Celestin, L. R. (1967). Gastrin-like effects of cholecystokinin-pancreozymin. *Nature (Lond.),* **215,** 763

39. Murat, J. E. and White, T. T. (1966). Stimulation of gastric secretion by commercial cholecystokinin extracts. *Proc. Soc. Exp. Biol. Med.,* **123,** 593

40. Magee, D. F. and Nakamura, M. (1966). Action of pancreozymin preparations on gastric secretion. *Nature (Lond.),* **212,** 1487

41. Grossman, M. I. (1970). Gastrin, cholecystokinin, and secretin act on one receptor. *Lancet,* **i,** 1088

42. Kimball, C. P. and Murlin, J. R. (1924). Aqueous extracts of pancreas. III. Some precipitation reactions with insulin. *J. Biol. Chem.,* **58,** 337

43. Collens, W. S. and Murlin, J. R. (1929). Hyperglycaemia following the portal injection of insulin. *Proc. Soc. Exp. Biol. Med.,* **26,** 485

44. Staub, A., Sinn, L. and Behrens, O. K. (1955). Purification and crystallisation of glucagon. *J. Biol. Chem.* **214,** 619

45. Wunsch, E. (1967). Total synthesis of the pancreatic hormone glucagon. *Z. Naturforsch. (B),* **22,** 1269

46. Baum, J., Simons, B. F., Jr., Unger, R. H. and Madison, L. L. (1962). Localisation of glucagon in the alpha cells in the pancreas islet by immunofluorescent techniques. *Diabetes,* **11,** 371

47. Sokal, J. E., Sarcione, E. J. and Henderson, A. M. (1964). Relative potency of glucagon and epinephrine as hepatic glycogenolytic agents. Studies with the isolated perfused liver. *Endocrinology,* **74,** 930

48. Sokal, J. E. (1966). Effect of glucagon on gluconeogenesis on the isolated perfused rat liver. *Endocrinology,* **78,** 538

49. Hagen, J. H. (1961). Effect of glucagon on the metabolism of adipose tissue. *J. Biol. Chem.,* **236,** 1023

50. Samols, E., Marri, G. and Marks, V. (1965). Promotion of insulin secretion by glucagon. *Lancet,* **ii,** 415

51. Crockford, P. M., Porte, D. Jr., Wood, F. C. and Williams, R. H. (1966). Effect of glucagon on serum insulin, plasma glucose and FFA in man. *Metabolism,* **15,** 114

52. Pohl, S. L., Birnbaumer, L. and Rodbell, M. (1969). Glucagon-sensitive adenylcyclase in plasma membrane of hepatic parenchymal cell. *Science,* **164,** 566

53. Samols, E., Tyler, J., Marri, G. and Marks, V. (1965). Stimulation of glucagon secretion by oral glucose. *Lancet,* **ii,** 1257

54. Samols, E., Tyler, J., Megyesi, C. and Marks, V. (1966). Immunochemical glucagon in human pancreas, gut and plasma. *Lancet,* **ii,** 727

55. Kenny, A. J. and Say, R. R. (1962). Glucagon-like activity extractable from the gastrointestinal tract of man and other animals. *J. Endocrinol.,* **25,** 1

56. Unger, R. H., Ketterer, H. and Eisentraut, A. M. (1966). Distribution of immunoassayable glucagon in gastrointestinal tissues. *Metabolism,* **15,** 865

57. Buchanan, K. D., Vance J. E., Aoki, T. and Williams, R. H. (1967). Rise in serum immunoreactive glucagon after intrajejunal glucose in pancreatectomised dogs. *Proc. Soc. Exp. Biol. Med.,* **126,** 813

58. Buchanan, K. D. (1969). The role of glucagon in health and disease. *MD thesis,* University of Glasgow

59. Valverde, I., Rigopoulou, D. Marco, J., Faloona, G. R. and Unger, R. H. (1970). Characterisation of glucagon-like immunoreactivity (GLI). *Diabetes,* **19,** 614

60. Valverde, I., Rogopoulou, D., Marco, J., Faloona, G. R. and Unger, R. H. (1970). Molecular size of extractable glucagon and glucagon-like immunoreactivity (GLI) in plasma. *Diabetes,* **19,** 624

61. Murphy, R. F., Buchanan, K. D. and Elmore, D. T. (1973). Isolation of glucagon-like immunoreactivity of gut by affinity chromatography on anti-glucagon antibodies coupled to Sepharose 4B. *Biochin. Biophys. Acta,* **303,** 118

62. Heding, L. G. (1971). Radioimmunological determination of pancreatic and gut glucagon in plasma. *Diabetologia,* **7,** 10

63. Buchanan, K. D., Flanagan, R. W. J., Murphy, R. F., O'Connor, F. A. and Shahidullah, M. (1974). Immunological characterisation of glucagon-like immunoreactivity. *Europ. J. Clin. Invest.* (abstract)

64. Flanagan, R. W. J., Buchanan, K. D. and Murphy, R. F. (1974). Specificity of antibodies in radioimmunoassay of glucagon. *Diabetologia,* **10,** 365

65. Jorpes, J. E., Mutt, V., Magnusson, S. and Steele, B. B. (1962). Amino acid composition and N-terminal amino acid sequences of porcine secretin. *Biochem. Biophys. Res. Commun.,* **9,** 275

66. Buchanan, K. D., Vance, J. E. and Williams, R. H. (1964). Insulin and glucagon release from isolated islets of Langerhans. Effect of enteric factors. *Diabetes,* **18,** 381

67. Jorpes. J. E. and Mutt, V. (1954). *Archiv. Kerm.,* **7,** 553

68. Wormsley, K. G. and Grossman, M. I. (1964). Inhibition of gastric acid secretion by secretin and endogenous acid in the duodenum. *Gastroenterology,* **47,** 72

69. Jorpes, J. E., Mutt, V., Jonson, G., Thulin, L. and Sundman, L. (1965). The influence of secretin and cholecystokinin on bile flow. In: *The Biliary System,* 283 (editor; W. Taylor) (Philadelphia: F. A. Davis Company)

70. Brown, J. C., Mutt, V. and Pederson, V. (1970). Further purification of a polypeptide demonstrating enterogastrone activity. *J. Physiol.,* **209,** 57

71. Brown, J. C. and Dryburgh, J. R. (1971). A gastric inhibitory polypeptide. II The complete amino acid sequence. *Can. J. Biochem.,* **49,** 867

72. Pederson, R. A. and Brown, J. C. (1972). Inhibition of histamine-, pentagastrin- and insulin-stimulated canine gastric secretion by pure 'gastric inhibitory polypeptide'. *Gastroenterology,* **62,** 393

73. Said, S. I. and Mutt, V. (1972). Isolation from porcine-intestinal wall of a vasoactive octasapeptide related to secretin and glucagon. *J. Biochem.,* **28,** 199

74. Brown, J. C. (1973). Motolin, a gastric motor activity stimulating polypeptide: the complete amino aicd sequence. *Can. J. Biochem.,* **51,** 533

75. Unger, R. H., Ketterer, H., Eisentraut, A. and Dupre, J. (1966). Effect of secretin on insulin secretion. *Lancet,* **ii,** 24

76. Jarrett, R. J. and Cohen, N. M. (1967). Intestinal hormones and plasma-insulin. Some observations of glucagon, secretin and gastrin. *Lancet,* **ii,** 861

77. Buchanan, K. D., Vance, J. E., Morgan, A. and Williams, R. H. (1968). Effect of pancreozymin on insulin and glucagon levels in blood and bile. *Amer. J. Physiol.,* **215,** 1293
78. Hinz, M., Katsilambros, N., Schweizer, B., Raptis, S. and Pfeiffer, E. F. (1971). The role of the exocrine pancreas in the stimulation of insulin secretion by the intestinal hormones. I. Effect of pancreozymin, secretin, gastrin, pentapeptide and of glucagon upon insulin secretion of isolated islets of rat pancreas. *Diabetologia,* **7,** 1
79. Young, J. D., Lazarus, L., Chisholm, D. J. and Atkinson, F. F. V. (1968). Radio-immunoassay of secretin in human serum. *J. Nucl. Med.,* **9,** 641
80. Chisholm, D. J., Young. J. D. and Lazarus, L. (1969). The gastrointestinal stimulus to insulin release. I, Secretin. *J. Clin. Invest.,* **48,** 1453
81. Chisholm, D. J., Kraegen, E. W. and Young, J. D. and Lazarus, L. (1971), Comparison of secretin response to oral and intraduodenal or intravenous glucose administration *Horm. Metab. Res.,* **3,** 180
82. Buchanan, K. D., Teale, J. D., Harper, G., Hayes, J. R. and Trimble, E. R. (1973). Plasma secretin assay in man. *Clin. Sci. Molec. Med.,* **45,** 13P
83. Buchanan, K. D., Teale, J. D., Harper, G. (1972). Antibodies to secretin using un-conjugated natural and synthetic secretins. *Horm. Metab. Res.,* **4,** 57
84. Sum, P. T. and Preshaw, R. M. (1967). Intraduodenal glucose infusion and pancreatic secretion in man. *Lancet,* **ii,** 340
85. Wang, C. C. and Grossman, M. I. (1951). Physiological determination of release of secretin and pancreozymin from intestine of dogs with transplanted pancreas. *Amer. J. Physiol.* **164,** 527
86. Dupre, J. and Beck, J. C. (1966). Stimulation of release of insulin by an extract of intestinal mucosa. *Diabetes,* **15,** 555
87. Unger, R. H., Ketterer, H., Dupre, J. and Eisentraut, A. M. (1967). The effects of secretin, pancreozymin and gastrin on insulin and glucagon secretion in anaesthetised dogs. *J. Clin. Invest.,* **46,** 630
88. Fussgänger, R. D., Straub, K., Goberna, R., Jaros, P., Schröder, K. E., Raptis, S. and Pfeiffer, E. F. (1969). *Horm. Metab. Res.,* **1,** 224
89. Young, J. D., Lazarus, L. and Chisholm, D. J. (1968). Secretin and pancreozymin-cholecystokinin after glucose. *Lancet,* **ii,** 914
90. Ohneda, A., Parada, E., Eisentraut, A. and Unger, R. H. (1968). Characterisation of response of circulating glucagon to intraduodenal and intravenous administration of amino acids. *J. Clin Invest.,* **47,** 2305
91. Fajans, S. S., Floyd, J. C., Knopf, R. F. and Conn, J. W. (1967). Effect of amino acids and proteins on insulin secretion in man. *Recent Progr. Horm. Res.,* **23,** 617
92. Unger, R. H. and Eisentraut, A. M. (1969). Entero-insular axis. *Arch. Int. Med.,* **123,** 261
93. Hayes, J. R., Ardill, J., Trimble, E. R. and Buchanan, K. D. (1973). Gastrin and the pancreas. *Europ. J. Clin. Invest.,* **3,** 236
94. Buchanan, K. D. (1973). Studies on the pancreatic enteric hormones. *PhD thesis,* The Queen's University of Belfast
95. Ardill, J. (1974). Personal communication
96. Turner, D. S. and McIntyre, N. (1966). Stimulation by glucagon of insulin release from rabbit pancreas *in vitro.* Lancet, **i,** 351
97. Devrim, S. and Recant, L. (1966). Effect of glucagon on insulin release *in vitro.* *Lancet,* **ii,** 1227

98. Ketterer, H., Eisentraut, A. M. and Unger, R. H. (1967). Effect upon insulin secretion physiologic doses of glucagon administered via the portal vein. *Diabetes*, **16**, 283

99. Dinstl, K. and Williams, R. H. (1969). Effect of blood glucose on glucagon secretion in anaesthetised dogs. *Diabetes*, **18**, 11

100. Brown, J. C. (1974). Physiology and pathophysiology of gastric inhibitory polypeptide (GIP) and physiology of motolin. *Proceedings of the 80th Congress of the German Society for Internal Medicine* (Wiesbaden; April, 1974)

101. Ross, S. A., Brown, J. C. and Dupre, J. (1974). Effects of gastric inhibitory polypeptide on endocrine pancreas in normal and diabetic subjects. *Diabetologia*, **10**, 384

102. Turner, D. S. (1974). Personal communication

103. Turner, D. S. (1969). Intestinal hormones and insulin release: *in vitro* studies using rabbit pancreas. *Horm. Metab. Res.*, **1**, 168

104. Moody, A. J., Markusson, J., Schaich-Fries, A., Steenstrup, C., Sundby, F., Malaisse, W. and Malaisse-Lagae, F. (1970). The insulin-releasing activities of extracts of pork intestine. *Diabetologia*, **6**, 135

105. Turner, D. S. and Marks, V. (1972). Enhancement of glucose-stimulated insulin release by an intestinal polypeptide in rats. *Lancet*, **i**, 1095

106. Deckert, T., Lauridsen, U. B., Madsen, S. N. and Mogensen, P. (1972). Insulin response to glucose, tolbutamide, secretin and isoprenaline in maturity-onset diabetes mellitus. *Danish Med. Bull.*, **19**, 222

107. Hindberg, I., Enk, B. and Persson, I. (1970). Insulin stimulation by secretin in diabetics. Effects of repeated and varied doses.

108. Raptis, S., Schroder, K. E., Faulhaber, J. D. and Pfeiffer, E. P. (1970). Stimulation of insulin secretion be secretin in diabetics. *Germ. Med. Mth.*, **15**, 206

109. Melani, F., Lawecki, J., Bartelt, K. M. and Pfeiffer, E. F. (1967). Insulinspiegel bei Stoffwechselgesunden, Fettsüchtigen und Diabetikern nach intravenöser Gake von Glukose, Tolbutamid und Glucagon. *Diabetologia*, **3**, 422

110. Simpson, R. G., Benedetti, A., Grodsky, G. M., Karam, J. H. and Forsham, P. H. (1966). Stimulation of insulin release by glucagon in non-insulin dependent diabetics. *Metabolism*, **15**, 1046

111. Schröder, K. E., Raptis, S., Faulhaber, J. D., Fussgänger, R. D., Straub, K. and Pfeiffer, E. F. (1970). Differences in the alpha- beta-cytotropic effect of pancreozymin in normal and diabetic subjects (abstract). *Diabetologia*, **6**, 81

112. Deckert, T., Kolendorf, K., Persson, I. and Worning, H. (1972). Insulin secretion after tolbutamide and after secretin in patients with pancreatic diseases. *Acta Med. Scand.*, **192**, 465

113. Raptis, S., Rau, R. M., Hartmann, W., Clodi, P. H. and Pfeiffer, E. F. (1970). Effect of secretin and pancreozymin on insulin secretion in patients with exocrine pancreatic insufficiency. *Diabetologia*, **6**, 61

114. Goterne, R., Fussgänger, R. D., Raptis, S., Telib, M. and Pfeiffer, E. F. (1971). The role of the exocrine pancreas in the stimulation of insulin secretion by intestinal hormones. *Diabetologia*, **7**, 68

115. Guidoix-Grassi, L. and Felber, J. P. (1968). Effect of secretin on insulin release by rat pancreas (abstract). *Diabetologia*, **4**, 386

116. Pfeiffer, E. F., Fussgänger, R., Hinz, M., Raptis, S., Schleyer, M. and Straub, K. (1971).Extra-pancreatic hormones and insulin secretion. In: *Diabetes, Proceedings of the 7th Congress of the International Diabetes Federation* Medica Excerpta (Buenos Aires)

117. Lerner, R. L. and Porte, D. (1972).Studies of secretin-stimulated insulin responses in man. *J. Clin. Invest.*, **51**, 2205

118. Aguilar-Parada, E., Eisentraut, A. M. and Unger, R. H. (1969).Pancreatic glucagon secretion in normal and diabetic subjects. *Amer. J. Med. Sci.*, **257**, 415

119. Unger, R. H., Aguilar-Parada, E., Müller, W. A. and Eisentraut, A. M. (1970). Studies of pancreatic alpha cell function in normal and diabetic subjects. *J. Clin. Invest.*, **49**, 837

120. Wise, J. K., Hendler, R. and Felig, P. (1973).Evaluation of alpha cell function by infusion of alanine in normal, diabetic and obese subjects. *New Engl. J. Med.*, **288**, 487

121. Buchanan, K. D. and McCarroll, A. M. (1972).Abnormalities of glucagon metabolism in untreated diabetes mellitus. *Lancet*, **ii**, 1394

122. Buchanan, K. D. and Mawhinney, W. A. A. (1973). Glucagon release from isolated pancreas in streptozotocin-treated rats. *Diabetes*, **22**, 797

123. Buchanan, K. D. and Mawhinney, W. A. A. (1973). Insulin control of glucagon release from insulin-deficient rat islets. *Diabetes*, **22**, 801

124. Trimble, E. R., Montgomery, D. A. D. and Hadden, D. R. (1974).Change in pattern of glucagon response during OGTT before and six months after commencement of intensive dietary management of diabetes. *Diabetologia*, **10**, 389

125. Ardill, J., Montgomery, D. A. D. and Hadden, D. R. (1974).Gastrin in diabetes. *Diabetologia*, **10**, 357

126. Pek, S. (1974).Personal communication

127. Williams, R. H. (1968). *Textbook of Endocrinology* (Philadelphia: W. B. Saunders)

128. Somogyi, M. (1959).Diabetogenic effect of hyperinsulinism. *Am. J. Med.* **26**, 169

129. Verner, J. V. and Morrison, A. R. (1958).Islet cell tumor and a syndrome of refractory watery diarrhoea and hypokalaemia. *Am J. Med.*, **25**, 374

130. Verner, J. V. and Morrison, A. B. (1968).Clinical syndromes associated with non-insulin producing tumours of the pancreatic islets. In: *Non-Insulin Producing Tumors of the Pancreas, International Symposium at Erlandgen.* 165 (Stuttgart: George Thieme Verlag)

131. Horton, R., Kane, J. P., Biglieri, E. G., Grodsky, G. M. and Forsham, P. H. (1963). Carbohydrate metabolism in primary aldosteronism (abstract). *Clin. Res.*, **11**, 83

132. Sagild, U., Anderson, V. and Andreason, P. B. (1961).Glucose tolerance and insulin responsiveness in experimental potassium depletion. *Acta Med. Scand.*, **169**, 243

133. Elias, E., Polak, J. M., Bloom, S. R., Pearse, A. G. E., Welbourn, R. B., Booth, C. C., Kuzio, M. and Brown, J. C. (1972). Pancreatic cholera due to production of gastrin inhibitory polypeptide. *Lancet*, **ii**, 791

134. Bloom, S. R., Polak, J. M. and Pearse, A. G. E. (1973).Vasoactive intestinal peptide and water-diarrhoea syndrome. *Lancet*, **ii**, 14

135. Said, S. I. and Mutt, V. (1970).Polypeptide with broad biological activity: isolation from small intestine. *Science*, **169**, 1217

136. McGavran, M. H., Unger, R. H., Recant, L., Polk, H. C., Kilo, C. and Levin, M. E. (1966). A glucagon-secreting alpha-cell carcinoma of the pancreas. *New Engl. J. Med.*, **274**, 408

137. Church, R. E. and Crane, W. A. J. (1967). A cutaneous syndrome associated with islet-cell carcinoma of the pancreas. *Brit. J. Dermat.*, **74**, 284

138. Becker, S. W., Kahn, D. and Rothmann, S. (1942). Cutaneous manifestations of internal malignant tumors. *Arch. Dermatol.*, **45**, 1069

139. Yoshinaga, T., Okuno, G., Yoshitake, S., Tsugii, T. and Nishikawa, M. (1966). Pancreatic A-cell tumour associated with severe diabetes mellitus. *Diabetes*, **15**, 709

140. Mallinson, C. N., Bloom, S. R., Warin, A. P., Salmon, P. R. and Cox, B. (1974). A glucagonoma syndrome. *Lancet*, **ii**, 1

141. Gleeson, M. H., Bloom, S. R., Polak, J. M., Henry, K., Dowling, R. H. and Pearse, A. G. E. (1971). Endocrine tumour in kidney affecting small bowel structure, motility and absorptive function. *Gut*, **12**, 773

142. Bloom, S. R. (1972). An enteroglucagon tumour. *Gut*, **13**, 520

143. Rabinowitch, I. M. (1949). Achlorhydria and its clinical significance in diabetes mellitus. *Amer. J. Dig. Dis.*, **16**, 322

144. Irvine, W. J., Clarke, B. F., Scarth, L., Cullen, D. R. and Duncan, L. J. P. (1970). Thyroid and gastric autoimmunity in patients with diabetes mellitus. *Lancet*, **ii**, 163

145. Dotevall, G. (1959). Incidence of peptic ulcer in diabetes mellitus. *Acta Med. Scand.*, **164**. 463

146. Wilkinson, J. F. (1937). Pernicious anaemia and diabetes mellitus. *Brit. Med.J.*, **2**, 723

147. Wilkinson, J. F. (1963). Diabetes mellitus and pernicious anaemia. *Brit. Med. J.*, **1**, 676

148. Falta, W. (1936), *Die Zuckerkrankheit*. 79. (Berlin: Urban & Schwaenberg)

149. Katsch, G. (1926), Handbook of Internal Medicine, 2nd edition.
editors G. von Bergmann and R. Stachelm. (Berlin: J. Springer).

150. Kassander, P. (1958). A symptomatic gastric retention in diabetics (gastroparesis diabeticorum). *Ann. Inter. Med.*, **48**, 797

151. Malins, J. (1968). *Clinical Diabetes Mellitus*. (London: Eyre & Spottiswoode)

152. Bridewell, T. and Whitehouse, F. W. (1961). Peroral jejunal biopsy in a patient with diabetic diarrhoea. *Diabetes*, **10**, 58

153. Buchan, J. D. (1962). Jejunal biopsy in diabetic steatorrhoea. *Gastroenterology*, **42**, 193

154. Malins, J. M. and French, J. M. (1957). Diabetic diarrhoea. *Quart. J. Med.*, **26**, 467

155. Baron, J. H. and Nabarro, J. D. N. (1973). Pancreatic exocrine function in maturity-onset diabetes mellitus. *Brit. Med. J.*, **4**, 25

156. Aktan, H. S. and Klotz, A. P. (1959). Fat absorption and pancreatic function in diabetes mellitus. *Ann. Int. Med.*, **49**, 820

157. Sun, D. C. H. and Shay, H. (1961). An evaluation of the starch tolerance test in pancreatic insufficiency. *Gastroenterology*, **40**, 379

158. Anderson, M. F., Davison, S. H. H., Dick. A. P., Hales, C. N. and Owens, J. (1970). Plasma insulin in pancreatic disease. *Gut*, **11**, 524

159. Leszt, L. and Vogel, H. (1946). Resorption of glucose from the small intestine of alloxan-diabetic rats. *Nature (Lond.)*, **157**, 551

160. Paul, F. and Drury, D. R. (1942). The rate of glucose absorption from the intestine of diabetic rats. *Amer. J. Physiol.*, **137**, 242

161. Dubois, R. S. and Roy, C. C. (1969). Insulin-stimulated transport of 3-0-methyl glucose across the rat jejunum. *Proc. Soc. Exp. Biol. Med.,* **130,** 931

162. Love, A. H. G. and Canavan, D. A. (1968). Effects of insulin on intestinal glucose absorption. *Lancet,* **ii,** 1325

163. Teale, J. D. and Love, A. H. G. (1972). The effects of tolbutamide and glibenclamide on intestinal glucose absorption. *Biochem. Pharmacol.,* **21,** 1839

164. Botros, M., Selim, R., Ghoneim, Kh., Ghareeb, A. and Wahba, N. (1974). Insulinotropic drugs. Effect of intestinal absorption of sugar. *Diabetes,* **23,** 112

165. Teale, J. D. and Love, A. H. G. (1973). Effect of glibenclamide on sugar transport by fed, starved and diabetic rat small intestine. *Biochem. Pharmacol.,* **22,** 997

166. Teale, J. D., Love, A. H. G. and Buchanan, K. D. Unplublished observations

167. Czyzyk, A., Tawecki, J., Sadowski, J., Ponikowska, I. and Szczepanik, A. (1968). Effect of biguanides on intestinal absorption of glucose. *Diabetes,* **17,** 492

168. Hollobaugh, S. L., Bhaskar Rao, M. and Kruger, F. A. (1970). Studies on the site and mechanism of action of phenformin. *Diabetes,* **19,** 45

169. Czyzyk, A., Szadkowski, M., Rogata, H. and Lawecki, J. (1973). Effect of phenformin on the exocrine function of the pancreas and on insulin secretion after intraduodenal infusion of HCl and intravenous injection of secretin. *Diabetes,* **22,** 932

170. Creutzfeldt, W. (1973). Gastrointestinal hormones and insulin secretion. *New Engl. J. Med.,* **288,** 1238

171. Vinik, A. I., Kalk, W. J., Keller, P., Beaumont, P. and Jackson, W. P. U. (1973). Overactivity of the entero-insular axis in maturity onset diabetes. *Lancet,* **ii,** 182

172. Gerich, J. E., Langlois, M., Noacco, C., Karam, J. H. and Forsham, P. H. (1973). Lack of glucagon response to hypoglycaemia in diabetes: evidence for an intrinsic pancreatic alpha cell defect. *Science,* **182,** 171

4

THE LIVER IN GLUCOSE HOMEOSTASIS IN NORMAL MAN AND IN DIABETES

Philip Felig

The central role of the liver in glucose homeostasis has been recognised since the classic studies of Claude Bernard. In a number of ways the liver occupies a unique position in carbohydrate regulation: (a) it is an organ of glucose production as well as glucose consumption; (b) it is exposed to insulin concentrations in portal venous blood which are 3–10-fold greater than those in the systemic circulation[1]; (c) it is the sole site of the glucoregulatory action of glucagon; and (d) absorbed hexoses reach the liver before being delivered to muscle and adipose tissue. It is the purpose of this chapter to review recent studies which have delineated the role of hepatic glucose balance in blood glucose regulation in normal man, in diabetes, in obesity and in liver disease. This discussion is not meant to be an exhaustive survey of the literature, but will tend to focus on studies undertaken in the author's laboratory in human subjects. For a detailed review of earlier studies on insulin and the liver involving experimental animals, the reader is referred to the excellent review by Madison[2].

NORMAL MAN

The postabsorptive state

The period following an overnight fast and preceding the ingestion of the breakfast meal is generally referred to as the postabsorptive state. At this point in time, the concentrations of hormones (insulin and glucagon) and substrates (glucose, amino acids, fatty acids) which were altered by meal ingestion during the preceding day have returned to baseline concentrations. Although the postabsorptive state represents a relatively unsteady condition, it nevertheless is a useful reference point since, under usual circumstances, it represents the period of metabolic transition from the fed to the fasted condition.

After an overnight fast, the reduction in insulin concentrations to basal levels (10–20 μU/ml) results in virtually total cessation of glucose uptake by insulin-dependent tissues such as resting muscle, adipose tissue and liver. Glucose uptake continues, however, by non-insulin-dependent tissues such as the brain, the formed elements of the blood and the renal medulla, at a combined rate of 2–3 mg/kg/min (150–200 mg/min). Maintenance of blood glucose homeostasis is achieved in this circumstance by the hepatic release of glucose at rates equal to those of tissue utilisation. The hepatic processes involved in the addition of glucose to the blood stream consist of glycogenolysis as well as gluconeogenesis. On the basis of studies employing splanchnic balances of gluconeogenic substrates[3] and the rate of disappearance in liver glycogen in biopsy samples[4], it is estimated that approximately 70–75% of hepatic glucose release is due to glycogenolysis and the remainder (25%) is a result of gluconeogenesis.

With regard to gluconeogenesis, the various precursor substrates available for conversion to glucose are the glycolytic intermediates lactate, pyruvate and glycerol, and the glycogenic amino acids. Re-cycling of glucose-derived lactate and pyruvate (Cori cycle) accounts for approximately 15% of glucose production[3 5], and an additional 2% is contributed by glycerol released from adipose tissue. The remainder of gluconeogenesis represents conversion of precursor amino acids, of which alanine makes the largest contribution, accounting for 6–12% of total glucose output, and 20–50% of the gluconeogenic component[6]. The predominance of alanine in the outflow of amino acids from muscle[7] and the evidence of its synthesis from glucose-derived pyruvate in muscle tissue[3 8 9] has led to the recognition of a glucose alanine cycle[3 7 10] analogous to the Cori cycle observed for lactate. These quantitative considerations of hepatic glucose production and peripheral glucose utilisation are summarised in Figure 4.1.

With regard to the hormonal factors responsible for hepatic glucose output in postabsorptive man it is clear that the fall in plasma insulin from postprandial to basal concentrations is crucial to the stimulation of hepatic glycogenolysis and the mobilisation of precursor amino acids and energy-yielding fatty acids necessary for gluconeogenesis[11]. The precise role of basal glucagon concentrations in this circumstance is less clear. The fall in postabsorptive blood glucose levels induced by inhibition of glucagon secretion by administration of somatostatin (growth hormone release-inhibitory factor) suggests that basal glucagon secretion is a necessary component in postabsorptive glucose homeostasis[12][13].

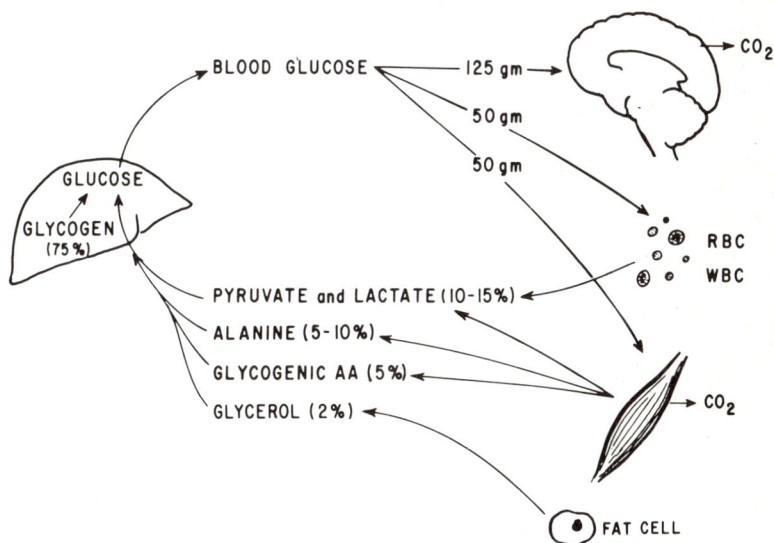

Figure 4.1 Glucose balance in normal man in the postabsorptive, overnight-fasted state.

Brief and prolonged starvation

As food intake is withheld beyond the overnight fasted condition, on-going glucose homeostasis is dependent on continuing hepatic glucose production. The overall rate of glucose production and the relative contributions from glycogenolysis and gluconeogenesis are influenced by the following considerations:

1. Total liver glycogen stores comprise no more than 70 grams after an overnight fast[4].
2. Protein-derived amino acids represent the sole precursors utilisable by mammalian liver for *de novo* glucose synthesis.
3. Dissolution of body protein beyond a critical threshold (30–50% decrease) generally results in death.

As a result of these constraints, the adaptation in hepatic glucoregulatory mechanisms to starvation is biphasic, characterised by an initial increase in gluconeogenesis and a subsequent fall in total glucose production[14]. The transition between these phases represents a gradual changeover, the initial response being most pronounced at about 3 days of fasting, and the secondary response being evident after 3–4 weeks of starvation.

Inasmuch as liver glycogen content is limited to 70 g[4] while glucose consumption (particularly by the brain) occurs at a rate of approximately 150 g/day (Figure 4.1), it is clear that hepatic glycogen stores are depleted within a 24-hour period of fasting. In fact, liver biopsy samples obtained after a 72-hour fast in normal humans reveal virtually total disappearance of glycogen which persists as fasting continues[4]. (This is in contrast to the situation in the rat, in which liver glycogen is repleted as fasting extends from 24 to 72 hours). To compensate for the lack of available glycogen, hepatic gluconeogenesis is stimulated as indicated by an increase in splanchnic uptake of alanine after a 48–72-hour fast[6]. This augmentation in gluconeogenesis is a consequence of hepatic as well as extra-hepatic events. Stimulation of hepatic gluconeogenic pathways is suggested by the demonstration of an increase in splanchnic fractional extraction of alanine[6]. The latter may be a result of progressive hypoinsulinaemia as well as the development of peak increments in plasma glucagon at 72 hours of starvation[15]. In addition, muscle output of alanine and other amino acids increases after a 72-hour fast[16]. Thus while total glucose release by the liver in 2–3-day fasted man is not greater than in the postabsorptive state, the relative contribution from gluconeogenosis is substantially increased to make up for the depletion of the glycogenolytic component. This dependence on gluconeogenesis from protein is reflected in the brisk rate of urinary nitrogen excretion early in starvation[17].

Inasmuch as death in starvation is a result of protein dissolution (leading to decreased respiratory muscle function, atelectasis and terminal pneumonia), rather than a consequence of hypoglycaemia, from a teleologic point of view, one would anticipate a progressive decrease in the rate of protein breakdown as starvation continues. That such an adaptation occurs has been recognised since the classical studies of Benedict revealed a fall in urinary nitrogen excretion from 10–12 g/day during the first week of fasting to 5 g or less after 4 weeks of starvation[18]. Since gluconeogenesis is dependent on protein-derived precursors, a simultaneous reduction in hepatic glucose production would be predicted. In fact, by 4–6 weeks of fasting, hepatic glucose production is reduced to 40 g/day, as compared to rates of 150–200 g/day in the postabsorptive state[17]. The rate-limiting step responsible for this decresase is extra-hepatic in location inasmuch as the fractional extraction of alanine by the liver is comparable to the postabsorptive state[6]. Furthermore, exogenous alanine is rapidly converted to glucose[19][20]. The diminution in glucose production in the face of intact hepatic gluconeogenic mechanisms is a result of a fall in endogenous alanine release from muscle[7] thereby decreasing the delivery to the liver of glucose precursors[6]. The reduction in alanine availability cannot be accounted for by altered hormonal secretion, but appears to be a consequence of starvation ketosis[21]. Recent studies have shown that an increase in circulating ketone acids results in a specific decline in plasma alanine and a reduction in protein catabolism in prolonged starvation[21].

The maintenance of glucose homeostasis in prolonged fasting in the face of reduced rates of hepatic gluconeogenesis is achieved by means of extra-hepatic alterations in glucose production as well as utilisation. After a 4–6-week fast, the kidney becomes an important source of glucose synthesis, accounting for approximately half of the total release of glucose into the bloodstream[16]. Of even greater quantitative importance in overall glucose balance, is a reduction in glucose utilisation to levels which are 50% below those observed in postabsorptive man. This is achieved by the utilisation of ketone acids by the brain as its main oxidative fuel in place of glucose[22]. Thus ketones subserve a dual role in starvation: (a) as substrate for the brain, thereby limiting glucose utilisation; (b) as a protein-sparing signal to muscle, thereby limiting alanine availability for gluconeogenesis.

Glucose ingestion
In the fasted condition, the liver contributes to glucose homeostasis by glycogenolysis and gluconeogenesis in response to hypoinsulinaemia and hyperglucagonaemia. An equally important role is played by the liver in the maintenance of normal glucose tolerance in response to carbohydrate ingestion. While the liver has long been recognised as a site of insulin action, the relative importance of hepatic as compared to peripheral muscle and adipose tissue in the disposal of an oral glucose load has only recently been quantitated[23].

Studying healthy subjects, Felig *et al.*[23] examined splanchnic glucose balance following the ingestion of 100 g of glucose. Since glucose is absorbed via the portal system, the extent to which an oral glucose load is available for uptake by peripheral tissues depends upon its escape from the splanchnic bed. As shown in Figure 4.2, following glucose ingestion, splanchnic glucose output increases rapidly, reaching values 2–3 times the basal rate within 15 minutes, and returning to baseline at 90 minutes. A secondary rise in splanchnic glucose output occurs at 150 minutes, coinciding with a secondary increment in blood glucose. The total amount of glucose escaping hepatic uptake and entering the systemic circulation over the 3–hour period of observation (determined by integrating the area under the splanchnic glucose output curve) was noted to be 40 g. Of even greater interest is the observation that splanchnic glucose delivery during this period exceeds the basal (postabsorptive) rate by only 15 g. As noted above, the basal rate of glucose utilisation represents consumption by non-

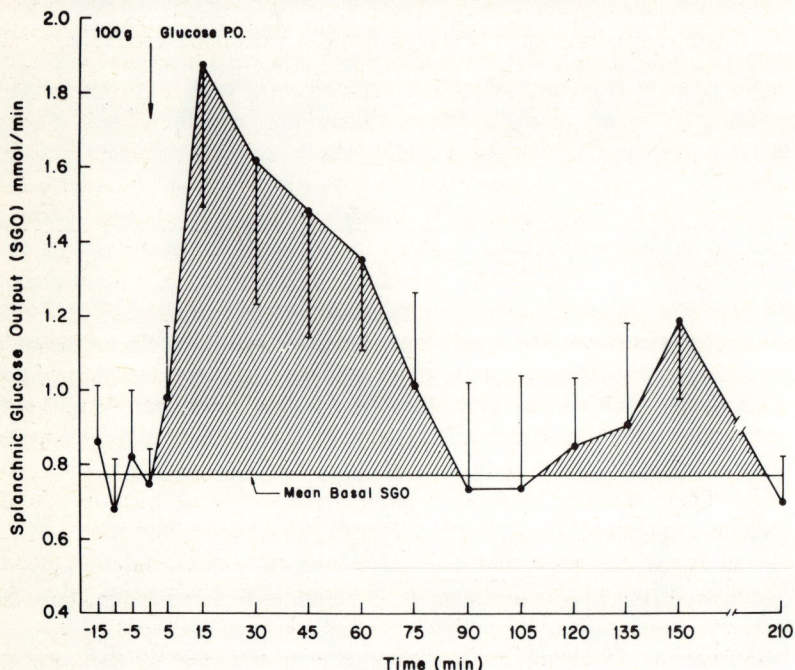

Figure 4.2 The effect of oral glucose ingestion on splanchnic glucose output. The horizontal line represents the mean splanchnic glucose output in the basal postabsorptive state. The hatched area indicates the increase in splanchnic glucose output above basal levels. Based on the data of Felig *et al.*[23].

insulin dependent tissues, particularly brain and blood cellular elements (Figure 4.1). Thus the total amount of glucose made available for disposal as increased (above basal) glucose utilisation by peripheral insulin-dependent tissues is only 15% of the ingested load.

From these observations, some approximations of the overall disposal of 100 g of ingested glucose may be made (Table 4.1). Over the 3-hour period after ingestion of an oral glucose load, less than 5% is retained within the glucose space, the remainder being taken up and/or metabolised by various tissues. As noted above, 15% is utilised by peripheral fat and muscle tissue

Table 4.1 Disposal of a 100 g oral glucose load in normal subjects*

Glucose remaining in glucose space (unmetabolised)	< 5%
Increased peripheral glucose utilisation (insulin-dependent)	15%
Hepatic disposal	
Glycogen-sparing (non-insulin dependent peripheral uptake)	25%
Hepatic retention	55%
Glycogen synthesis	
Triglyceride formation	
Glycolysis	

*Based on the data of Felig *et al.*[23].

along insulin-dependent pathways. An additional 25% escapes the splanchnic bed, but is utilised to meet the on-going obligate glucose needs of brain and other non-insulin dependent tissues. Since these tissues would otherwise have been dependent on hepatic glycogenolysis, this proportion of total glucose utilisation (25%) may be viewed as glycogen-sparing (Table 4.1). The remaining 55–60% of the glucose load is taken up by the liver where it is utilised for glycogen synthesis, triglyceride formation and (to a small extent) glycolysis[23]. It is thus apparent that as compared to the liver, adipose tissue and muscle represent minor sites of disposal of ingested glucose. Similar conclusions have been reached on the basis of less direct observations involving comparisons of the response to the oral and intravenous administration of glucose[24 25].

The importance of the liver in glucose homeostasis in the fed state is further underscored by its role in determining the height and shape of the glucose tolerance curve and in being the target site responsible for glucose-induced changes in lactate, pyruvate and alanine. Thus the peak rise in blood glucose induced by oral glucose is directly proportional to the increment in splanchnic glucose escape[23]. In addition, the secondary elevation in blood glucose often observed at 150 minutes coincides with a secondary rise in splanchnic glucose output.

With regard to lactate, pyruvate and alanine metabolism, it is of interest that for many years insulin-induced inhibition of gluconeogenesis was considered to be mediated via a reduction in peripheral output of gluconeogenic precursors rather than a result of a direct hepatic effect[26].

Observations on splanchnic balances of glucose precursors however, support a direct effect on the liver. Following oral[23] or intravenous[27] glucose administration, splanchnic uptake of lactate and pyruvate falls by 60–100%, or (in the case of pyruvate) may revert to a net output. Inasmuch as blood lactate and pyruvate levels rise after glucose administration[23] [27], a reduction in substrate presentation cannot be invoked to account for the decrease in hepatic utilisation of these glucose precursors. In fact primary inhibition of hepatic uptake may well account for the accumulation of lactate and pyruvate in blood induced by glucose inasmuch as peripheral output of these substrates is not enhanced by glucose ingestion[28]. In a like manner, the inhibition in gluconeogenesis from amino acids induced by hyperinsulinaemia and hypoglucagonaemia is due to decreased fractional extraction of alanine by the liver [23] [27] rather than a consequence of reduced peripheral output of this amino acid[29] (Figure 4.3). As in the case of lactate and pyruvate, blood alanine levels show a small but significant rise after glucose administration (Figure 4.3), which may be attributed to decreased hepatic uptake. It should be noted that glucose-induced insulin secretion is associated with a reduction in specific amino acid levels[30]. However this hypoaminoacidaemic effect involves primarily the branched chain amino acids (valine, leucine, isoleucine)[23] [27] [30] (Figure 4.3), which are not important endogenous glucose precursors[3] [6].

These observation on glucose metabolism in the fed state thus may be summarised as follows: (a) the liver is quantitatively the major site of disposal of an oral glucose load; (b) changes in blood lactate and pyruvate levels induced by glucose are a consequence of altered hepatic metabolism of these substrates; and (c) inhibition of gluconeogenesis from amino acids, lactate and pyruvate, associated with glucose-stimulated insulin secretion and hypoglucagonaemia is a result of diminished hepatic uptake of these substrates rather than a passive consequence of decreased peripheral output of glucose precursors.

Hepatic sensitivity to insulin
In evaluating the contribution of the liver to glucose homeostasis, it is of interest to determine (a) the relative sensitivity of hepatic and extra-hepatic tissues to small increments in circulating insulin; (b) the relative importance of insulin and glucagon in mediating glucose-induced changes in hepatic glucose balance; and (c) the relative responsiveness of glycogenolysis and gluconeogenesis to hormonal modulation.

Concerning tissue sensitivity to insulin, the following question may be posed: To what extent does diminished hepatic glucose output rather than

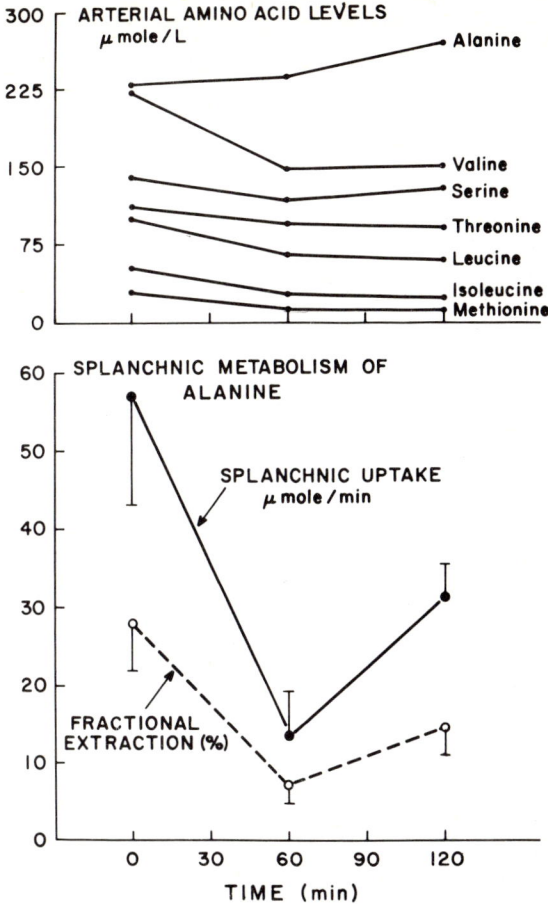

Figure 4.3 The effects of hyperinsulinaemia and hypoglucagonaemia induced by oral glucose ingestion on amino acid metabolism. *Upper Panel:* The rise in arterial alanine and the fall in the branched chain amino acids (valine, leucine and isoleucine) are statistically significant. *Lower Panel:* The decrease in splanchnic alanine uptake is a consequence of decreased fractional extraction of this amino acid. Based on the data of Felig *et al.*[23].

augmented peripheral glucose utilisation contribute to the overall effect of small elevations in serum insulin? Since the changes in serum insulin observed with ingestion of 100 g of glucose represent 5–10-fold increments above the basal concentrations[31], to examine the effects of small increments in serum insulin, normal subjects were studied during the infusion of glucose at a rate of 2 mg/kg/min (150 mg/min)[27 32]. This rate of glucose infusion results in increases in blood glucose of 15–20 mg/100 ml, and in elevations in serum insulin which are less than twice the basal levels; plasma glucagon levels are not changed by this infusion rate (Figure 4.4). Despite the small rise in insulin and the unchanged glucagon levels, a profound decline in splanchnic glucose output is observed (Figure 4.4). Glucose release from the liver is reduced 80–85% by the glucose infusion[27] [32]. In contrast to these effects on hepatic glucose balance, estimated glucose utilization (by peripheral tissues) is no greater than in the basal state (Figure 4.4). These observations indicate that small increments in endogenous insulin exert a profound effect on hepatic glucose balance,

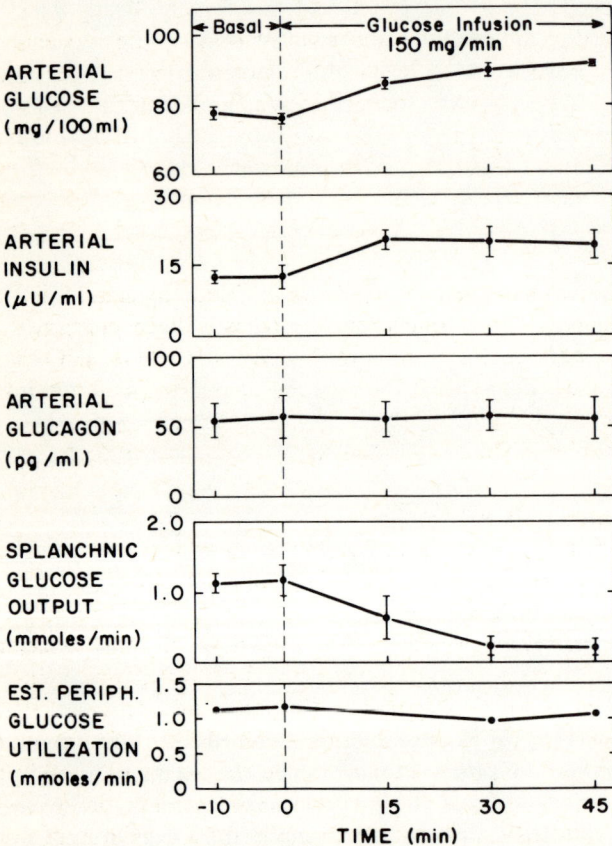

Figure 4.4 Hepatic sensitivity to insulin in normal man. Infusion of glucose at a rate (150 mg/min) causing no more than a 50–60% increase in insulin and no change in glucagon, results in an 80–85% inhibition of splanchnic glucose output. In contrast, peripheral glucose utilisation remains unchanged. Data of Felig *et al.*[27] [32] [42] [49].

yet fail to stimulate peripheral glucose uptake. The greater responsiveness of the liver as compared to fat and muscle tissues to small changes in insulin need not reflect an inherently greater sensitivity on the part of the hepatocyte to a given insulin concentration. Rather it may be a consequence of the higher ambient levels of endogenous insulin in portal as compared to peripheral blood[1]. It should also be noted that the alteration in hepatic glucose balance associated with small increments in serum insulin occurs in the absence of decreased glucagon secretion[32]. The available data thus indicate that insulin rather than glucagon is responsible for the fine modulation of glucose metabolism induced by small fluctuations in blood glucose and that the liver is the site of this insulin effect.

With regard to the relative responsiveness of hepatic glycogenolytic and gluconeogenic processes to insulin, it is noteworthy that the decline in splanchnic glucose output associated with small increments in insulin was not accompanied by changes in splanchnic uptake of lactate, pyruvate and alanine[27][32]. Interestingly the extent to which glucose output persisted (15–20% of basal levels) coincided with the rate of glucose production normally attributable to gluconeogenic precursor utilisation in postabsorptive man[3]. These observations thus suggest that glycogenolysis is more sensitive than gluconeogenesis to inhibition by small increments in insulin secretion[27]. This conclusion has recently been confirmed on the basis of observations on the response to high and low insulin doses using isotopically labelled alanine[33]. With greater increments in serum insulin (as occurs after ingestion of 100 g of glucose), gluconeogenesis as well as glycogenolysis is inhibited (see above).

Glucose homeostasis in exercise

In the resting state, glucose is a relatively unimportant fuel for muscle[34], the major source of energy being provided by oxidation of fatty acids[35]. During exercise, the total blood-borne fuel requirements of muscle increase markedly. In addition to augmented utilisation of fatty acids, glucose uptake and oxidation by muscle increased 5–7-fold during light exercise (400 kg-m/min; oxygen consumption 3–4 times basal) and rises 10–20-fold during moderate to severe exercise (800–1200 kg-m/min; oxygen consumption 6–10 times basal)[36]. The overall rate of total body glucose consumption rises 2–4-fold above basal levels depending on the severity of the work performed[36]. Exercise thus constitutes a condition of peripheral glucose need in which maintenance of glucose homeostasis depends on an increase in hepatic glucose output.

Studies involving hepatic venous catheterisation[36][37] as well as measurements of hepatic glycogen content[38] have documented increases in hepatic glucose release in response to exercise which reach values 5 times the basal rate of glucose ouput (Figure 4.5). In fact, over a wide range of work intensities, close agreement has been noted between the estimated rates of glucose uptake by exercising leg muscles and splanchnic glucose production[36]. Furthermore, although the kidney has the enzymatic capacity for glucose synthesis and is an important gluconeogenic organ during prolonged fasting[17], direct measurements of substrate balances during exercise have demonstrated no net renal glucose production[36]. Thus, hepatic glucose mobilisation represents the sole source of increased glucose delivery for consumption by exercising muscle.

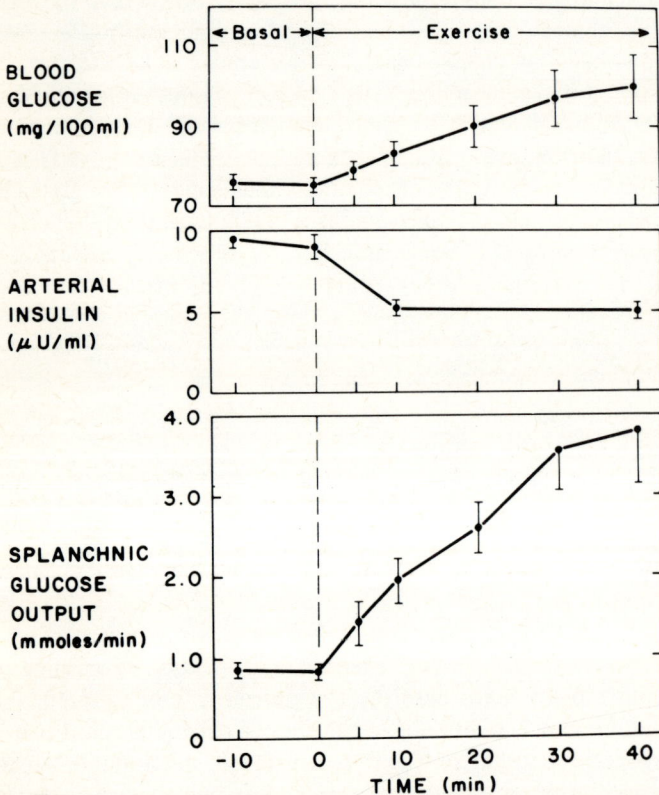

Figure 4.5 Stimulatory effects of exercise on splanchnic glucose output. Insulin levels are reduced by exercise[36 42].

The liver and glucose homeostasis

The intensity and duration of the exercise are the critical factors determining the absolute rate of glucose release and the relative contributions of glycogenolysis and gluconeogenesis. During exercise of mild intensity and short term duration (10–40 minutes), total glucose output rises 25–100% and the contribution from gluconeogenesis remains comparable to that which is observed in the basal state (i.e. 15–25%)[36]. During moderate to severe exercise (800–1200 kg-m/min), glucose release rises 100–400% and the contribution from gluconeogenesis decreases to 6–11[36]. This relative decrease in gluconeogenesis does not represent a fall in the absolute uptake of gluconeogenic precursors below basal resting levels. Rather it is indicative of a failure of precursor uptake to keep pace with the overall increase in hepatic glucose output. Although splanchnic fractional extraction and arterial concentrations of alanine[8], lactate and pyruvate[36] increase during exercise, the fall in splanchnic blood flow induced by moderate to severe exercise is of such magnitude as to preclude an increment in glucose precursor uptake. Thus the major source of hepatic glucose output during short term exercise, particularly at heavy work loads, is hepatic glycogenolysis. The extent of hepatic glycogen depletion during 40 minutes of severe exercise is approximately 18 g[36].

During more prolonged exercise the limitations in total hepatic glycogen stores necessitate a progressively increasing contribution from gluconeogenesis. It is estimated that after 4 hours of exercise, 55–60 grams of glycogen have been dissipated, representing $\frac{1}{4}$ of total hepatic glycogen stores[39]. Since very prolonged exercise (4 hours or more) can only be maintained at low work intensities (400–600 kg-m/min, 30–35% of maximal oxygen uptake), the marked diminution in splanchnic blood flow noted in severe exercise[36] is not observed in this circumstance, splanchnic blood flow remaining at basal levels[39]. As a result of an accompanying increase in splanchnic fractional extraction and in arterial levels of alanine, lactate and pyruvate, the splanchnic uptake of these glucose precursors increases 2–3-fold above basal levels. Since total glucose output rises by no more than 80%, the relative contribution from gluconeogenesis increases from resting levels of 20–25% to almost 50% after 4 hours of exercise[39]. Prolonged, mild exercise thus differs from the short term situation in that an increasing rather than decreasing contribution from gluconeogenesis is observed. With exercise extending for periods of 8 hours or more, virtually the entire hepatic glucose output is attributable to gluconeogenesis[40].

As to the factors responsible for augmented hepatic glucose output in exercise, a decrease in serum insulin is observed in short term[36] as well as prolonged exercise[39] (Figure 4.5). In view of the marked sensitivity of the liver to small changes in insulin[27] (see above), such a diminution in insulin would be expected to result in stimulation of hepatic glycogenolysis and gluconeogenesis. During heavy work intensities[41] and in very prolonged exercise[39], increases in plasma glucagon may contribute to hepatic glucose production as well. Recent studies suggest, however, that changes in circulating levels of insulin, glucagon, and glucose are not the sole (or primary) determinants of hepatic glucose production in exercise[42]. In healthy subjects given an infusion of glucose at a rate of 150 mg/min, before and during exercise, plasma insulin, glucagon and glucose concentrations were maintained at basal levels. Nevertheless, hepatic glucose production increased 2–3-fold[42]. Furthermore, infusion of insulin and maintenance of hyperinsulinaemia in excess of 100 μU/ml fails to inhibit exercise-induced stimulation of hepatic glucose output (Felig, P. and Wahren, J., unpublished data). Thus factors other than the prevailing concentration of insulin and glucagon contribute to the regulation of hepatic glucose output in exercise. Inasmuch as release of catecholamines is augmented in exercise, they may well influence hepatic glucose release in this circumstance.

DIABETES MELLITUS
In diabetes mellitus underutilization as well as overproduction of glucose by the liver are characteristic findings. Abnormalities in the net hepatic balance of glucose and glucose precursors may be observed in the fed as well as the fasted state depending on the severity of the metabolic defect. In this section, the role of the liver in diabetes will be considered in the context of the degree of clinical abnormality, namely glucose intolerance, fasting hyperglycaemia and diabetic ketoacidosis. The effect of diabetes on the hepatic response to exercise and the role of the liver in the future treatment of diabetes will also be considered.

Glucose intolerance
In the patient with a normal fasting blood glucose but an abnormal response to an oral glucose load, the postprandial hyperglycaemia is due, at least in part to a failure of hepatic glucose uptake. As noted above, the peak blood glucose concentration during an oral glucose tolerance test is directly proportional to the magnitude of splanchnic glucose escape[23]. Studies of peripheral glucose uptake have also shown that hepatic rather than peripheral utilisation determines the height and shape of the glucose tolerance curve[28]. Although direct quantitation of splanchnic glucose escape in response to oral glucose is available in normal subjects[23],

Table 4.2 Correlation between changes in insulin secretion and glucose uptake in mild diabetes

	Insulin secretion		Glucose uptake	
	Early	Late	Hepatic	Peripheral
Normal GGT	N1	N1	N1	N1
'Lag storage' GTT	→	N1 or ←	→	←
Mildly abnormal GTT	→	N1	→	N1
Moderately abnormal GTT	→	→	→	→

GTT = Glucose tolerance test
N1 = normal
← = increased
→ = decreased

comparative data have not been reported in diabetics. Indirect estimates however, suggest that in contrast to the 30–40 g of a 100-g oral load which enters the systemic circulation in normal man, in the diabetic, 50–75 g escapes hepatic uptake[25]. It should be recalled that of the 30–40 g of glucose which is delivered to peripheral tissues in normal man, only 15 g represents insulin-dependent above-basal glucose utilisation[23] (Table 4.1). Thus in the mild diabetic, an abnormal glucose tolerance curve would result from decreased hepatic glucose utilisation even in the face of normal rates of peripheral utilisation. That imbalances in hepatic and peripheral glucose uptake do in fact occur is suggested by the lag-storage glucose tolerance curve observed in some patients prior to the development of overt diabetes[43]. Such curves are characterised by unusually high blood glucose levels at 30/60 minutes (>160 mg/100 ml), but normal values at 2 and 3 hours[43]. The early hyperglycaemia probably reflects a reduction in hepatic uptake of ingested glucose resulting in greater delivery to the systemic circulation. The subsequent normalisation of blood sugar can be accounted for by a compensatory increase in peripheral glucose utilisation. Such an explanation would be consistent with the defect in insulin secretion observed in mild diabetics and in those individuals genetically predisposed to the development of diabetes. In such patients, it is the early insulin secretory burst which is defective[25 31 44 45], delayed hyperinsulinaemia often being observed.

Based on these considerations, progressive changes in glucose tolerance may be correlated with altered insulin secretion on the one hand, and decreased hepatic and/or peripheral glucose utilisation on the other hand (Table 4.2). By this formulation, in the mildest forms of glucose intolerance, decreased hepatic glucose uptake is the most sensitive index of a reduction in the early insulin secretory response to an oral glucose load. Peripheral glucose utilisation continues at normal rates in such patients as a result of an intact late secretory response to ingested glucose. With more severe insulin deficiency (early and late phase), hepatic and peripheral glucose uptake are reduced and more severe glucose intolerance ensues (Table 4.2). It should be noted that while a decrease in insulin secretion has been stressed as the primary pathogenic event in diabetes, failure of normal suppression of glucagon secretion by ingested glucose may contribute secondarily to the abnormal glucose tolerance curve[46].

Fasting hyperglycaemia
As the metabolic defect in diabetes becomes more manifest, hyperglycaemia characterises the fasted as well as the postprandial state.

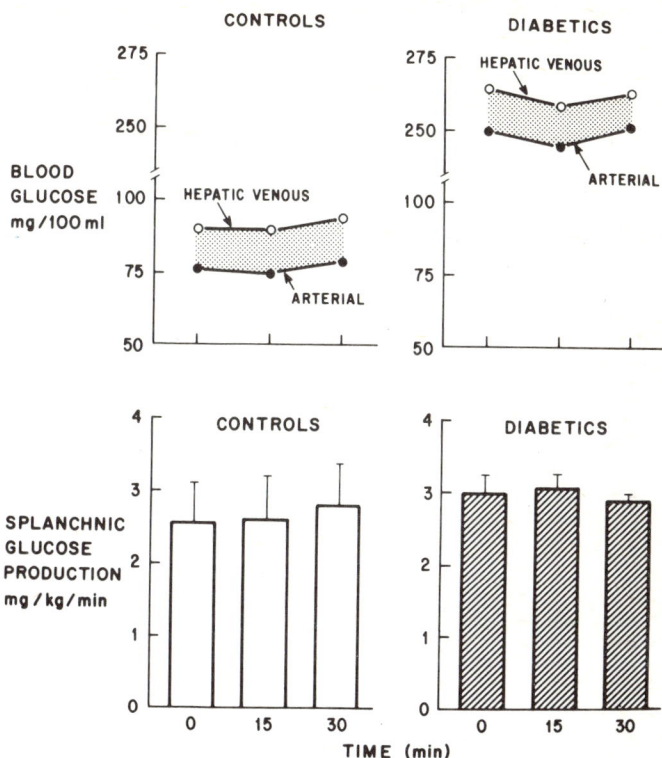

Figure 4.6 Blood glucose levels and splanchnic glucose production in normal controls and in diabetics with fasting hyperglycaemia. The absolute rate of splanchnic glucose production is not increased in the diabetics but is 'inappropriate' for the ambient blood glucose level[49].

Since the liver in the postabsorptive state normally releases glucose (see above), abnormalities in hepatic glucose balance in this circumstance must relate to alterations in hepatic glucose production rather than changes in glucose utilisation. Employing the hepatic venous catheter technique[47-49] or isotopic methods[50-52], a number of investigators have documented normal rates of hepatic glucose production in patients with fasting hyperglycaemia in the range of 150–350 mg/100 ml (Figure 4.6). On closer consideration, these seemingly normal rates of glucose production are clearly abnormal when examined in the context of the ambient blood glucose concentration and the relative contribution of gluconeogenesis to total glucose output.

As discussed above (see Hepatic sensitivity to insulin), in normal man a rise in blood glucose of 15–20 mg/100 ml is sufficient to inhibit hepatic glucose output by 80–85%[27] (Figure 4.4). In contrast, the diabetic with fasting hyperglycaemia releases glucose from the liver at "normal" rates of 150–250 mg/min despite an elevation in blood glucose of 100–200 mg/100ml (Figure 4.6). Presumably it is only in the face of the stimulatory effects of marked hyperglycaemia that β-cell secretion in such patients reaches "normal" basal levels (10–20 μU/ml, 1 Unit/hour) and "normal" rates of hepatic glucose output (rather than absolute overproduction) are achieved. Fasting hyperglycaemia thus may be viewed as an abnormality in the "glucostat" in which there is a relative rather than absolute overproduction of glucose, i.e. the rate of hepatic glucose release is inappropriate for the elevated blood glucose level.

The abnormality in hepatic glucose balance in fasting hyperglycaemia is even more apparent when the utilisation of gluconeogenic substrates is considered. As shown in Figure 4.7, the splanchnic uptake of glycogenic amino acids (particularly alanine) as well as lactate and pyruvate is twice that observed in normal subjects[49]. Consequently the overall contribution from gluconeogenesis to total hepatic glucose output is double the normal rate. Particularly noteworthy is the fact that this increase in

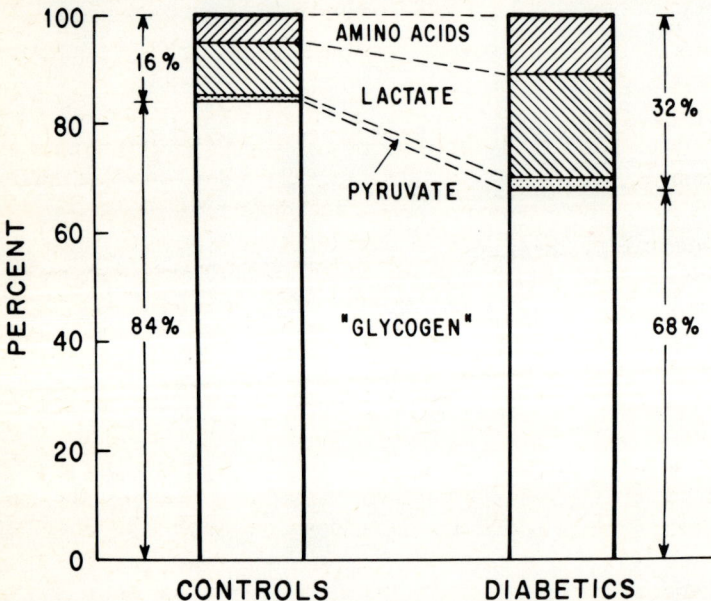

Figure 4.7 The contributions of gluconeogenic precursors (as indicated by net splanchnic balances) to total splanchnic glucose production in normal control subjects and diabetics[49]. The portion of total glucose output not accountable by precursor uptake is presumed to reflect glycogenolysis.

gluconeogenesis cannot be explained as a passive result of increased precursor supply; circulating levels of glucose precursors are either decreased (in the case of alanine and other glycogenic amino acids) or normal (lactate and pyruvate)[49]. On the other hand, splanchnic fractional extraction of glycogenic substrates is 2–3-fold higher in diabetic as compared to normal subjects. Thus the absolute increase in gluconeogenesis observed in the hyperglycaemic diabetic is a consequence of altered intra-hepatic events rather than a secondary result of increased delivery of glucose precursors[49]. These findings, implicating the liver as the target site of the gluconeogenic defect in insulin deficiency, are in keeping with the observations on insulin-induced inhibition of gluconeogenesis in normal man. As noted above, glucose-stimulated hyperinsulinaemia in healthy subjects results in an inhibition of splanchnic utilisation of glucose precursors despite an increase in circulating levels of these substrates[23][27].

Diabetic ketoacidosis

In its most florid state, diabetes is manifest as ketoacidosis, a condition characterised by marked hyperglycaemia and hyperketonaemia resulting in hyperosmolarity and metabolic acidosis. In such circumstances the liver occupies a key role with respect to the accumulation of blood glucose as well as ketones.

In patients with marked fasting hyperglycaemia (>400 mg/100 ml)[53] and/or ketoacidosis[54], the absolute rate of hepatic glucose production is two times or more the normal level. Since glucose utilisation in the fasting state is not insulin–dependent, severe accumulation of blood glucose is a reflection of the marked hepatic overproduction of glucose rather than a result of decreased glucose utilisation. Furthermore, direct balance studies indicate that the liver is the sole site of glucose production, since there is no net release of glucose from the kidney in diabetics (Wahren and Felig, unpublished data). Inasmuch as liver glycogen stores are limited to 70 g in the postabsorptive state, continued hepatic output of glucose at elevated rates in the face of anorexia or vomiting (which characterise ketoacidosis) can occur only as a result of augmented gluconeogenesis. As in patients with moderate degrees of fasting hyperglycaemia (see above), the increase in gluconeogenesis in diabetic ketoacidosis occurs in association with a reduction in circulating levels of glycogenic amino acids[55]. Primary stimulation of hepatic processes involved in the uptake and utilisation of glucose precursors thus characterises the severe diabetic state. In addition to insulin deficiency, an absolute increase in plasma glucagon may, in some cases, contribute to these abnormalities in hepatic glucose production[56].

Besides its importance in glucose synthesis, the liver is a locus of the metabolic defect in severe diabetes resulting in ketone overproduction. Increased hepatic ketogenesis in diabetes was for many years considered a passive consequence of augmented lipolysis and increased delivery of free fatty acids. More recent studies suggest that stimulation of β-oxidative pathways of fatty acid metabolism within the liver may be equally important[57]. Secifically enhanced activity of the acetyl-carnitine transferase reaction has been implicated as a key regulatory site in hepatic ketogenesis[58]. Thus with respect to increased production of ketones as well as glucose, intra-hepatic processes are stimulated in the severe diabetic state regardless of the level of precursor supply from peripheral tissues.

Response to exercise

In normal subjects, glucose utilisation during exercise is not insulin-dependent as reflected by augmented glucose consumption in association with a fall in serum insulin[36]. In keeping with these observations, glucose utilisation by muscle in exercising diabetics is comparable to that observed in healthy individuals[59 60]. On the other hand, with respect to glucose release as well as ketone production, the hepatic response to exercise in diabetics clearly differes from the normal population.

During short term (10–40 minutes), moderate to severe exercise (55–60% of maximal capacity), the total rate of hepatic glucose release in diabetics is comparable to controls (3–4 times resting, basal levels). However, the diabetics and normals differ with regard to the effect of exercise on gluconeogenesis. In the normal group the absolute rate of gluconeogenic precursor utilisation is unchanged from the resting state, resulting in a 50% or more decrease in the relative contribution from gluconeogenesis[36 60]. In contrast, in diabetics gluconeogenesis during exercise is increased 2–3-fold above resting levels[60]. This increase in gluconeogenic precursor uptake during short term exercise is a consequence of an increase in splanchnic fractional extraction as well as elevated arterial levels. The stimulative effect of short term exercise on gluconeogenesis in diabetes thus is comparable to that observed with long term exercise (4 hours or more) in healthy subjects in whom the gluconeogenic component rises[39], and suggests that diabetes accelerates the metabolic adaptation to exercise.

The effect of exercise on hepatic ketogenesis in diabetics depends on the extent of ketonaemia in the resting state. In patients with ketone acid levels below 1 mM, exercise causes no change in splanchnic output of ketones, while splanchnic uptake of free fatty acids (FFA) remains unchanged or falls[60]. In contrast, in patients with mild ketonaemia (2–4 mM), exercise results in a rise in arterial FFA, augmented splanchnic uptake of FFA, and a 2–3-fold increase in splanchnic output of ketone acids[60]. Thus with respect to ketogenesis (as well as gluconeogenesis), short term exercise in mildly ketotic diabetic subjects results in an intensification rather than an amelioration of the diabetic state. Nevertheless, during exercise blood ketone acid levels fail to rise above resting levels, as utilisation of ketone acids by exercising muscle keeps pace with augmented hepatic production[60].

THE LIVER IN THE FUTURE MANAGEMENT OF DIABETES

Insulin injection and diet have been the standard therapeutic modalities in the management of diabetes for over 50 years. Of late there has been an increasing awareness of the lack of success of these regimens with regard to the prevention of the microangiopathic and neuropathic complications of diabetes. Consequently interest has turned towards the possibility of developing a true cure for diabetes in the form of transplanted islets[61] or mechanical glucose-sensing insulin-delivery devices (artificial pancreas)[62]. As emphasized in the discussion above, the liver represents the major site of insulin action in normal man and the primary locus of the metabolic defect in the diabetic. Consequently it has recently been suggested that cure of the diabetic state may depend on restoration of normal portal-peripheral insulin gradients[63]. Such considerations thus may influence the site of implantation of transplanted islets (i.e. portal vein) and may limit the ultimate practicality of an artificial pancreas should a portal (rather than a peripheral) insulin injection site be necessary to truly normalise overall fuel metabolism in the diabetic.

OBESITY

It is well recognised that obesity has a profound influence on carbohydrate metabolism and insulin secretion. Hyperinsulinaemia[30][31], decreased responsiveness of fat[64] and muscle tissue[30][65] to insulin and an increased incidence of diabetes[66] have all been documented in obesity. Recent studies indicate that obesity also alters hepatic responsiveness to insulin[32]. The liver in the obese subject with normal glucose tolerance behaves similarly to that of the diabetic with fasting hyperglycaemia.

Measurements by splanchnic balance techniques and by isotope infusion demonstrate that total hepatic glucose output in obesity in the basal postabsorptive state is comparable to that observed in non-obese controls[32][67][68]. On the other hand, the relative contribution from

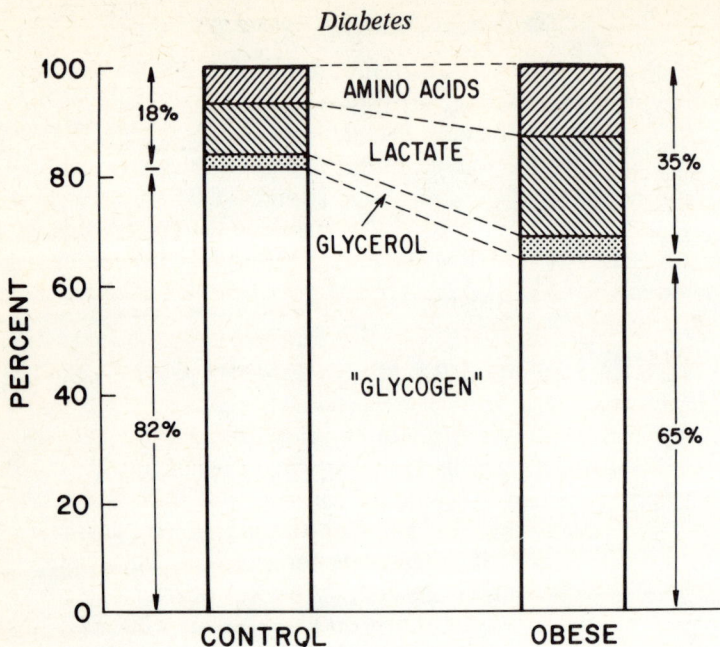

Figure 4.8 The increased contribution of gluconeogenesis to splanchnic glucose output in obesity. Based on the data of Felig *et al.*[32].

gluconeogenesis is increased by 70%[32]. As shown in Figure 4.8, the splanchnic uptake of lactate, glycerol and glycogenic amino acids (notably alanine) is increased in the obese group and can account for over 30% of total glucose output as compared to 19% in non-obese controls. The increase in alanine and lactate uptake in obesity is primarily due to an elevation in splanchnic fractional extraction of these substrates, a situation analogous to that observed in diabetes[49]. Studies with isotopically labelled alanine[69], lactate[70] and glycerol[71] also reveal augmented incorporation of these precursors into glucose in obesity. It is noteworthy however, that the increased hepatic avidity for and utilisation of glycogenic substrates which is strikingly similar to that observed in diabetes (Figure 4.7), occurs in obesity in the face of basal hyperinsulinaemia (serum insulin levels in obesity are 2–3 times greater than controls[30-32]). These findings thus suggest that the obese liver is resistant to the actions of insulin.

More direct evidence of altered hepatic responsiveness to insulin is demonstrable from studies examining the hepatic response to glucose infusion. When obese and non-obese subjects are given infusions of glucose (75–150 mg/min) resulting in comparable increments in serum insulin (50–60%) and blood glucose (10–15 mg/100 ml), a much smaller decline in

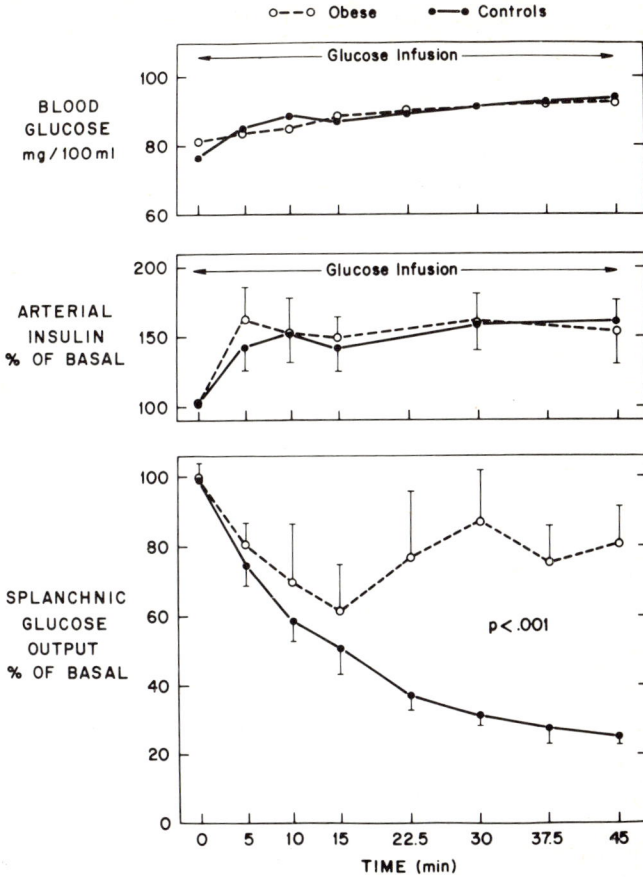

Figure 4.9 The splanchnic response to equivalent increments in plasma glucose and systemic insulin levels in obese and non-obese control subjects. Despite comparable increments in glucose and insulin, splanchnic glucose output fell no more than 40% in the obese group but declined 75–80% in the controls[32]. Glucagon levels were unchanged in both groups[32].

splanchnic glucose output is observed in the obese group (Figure 4.9)[32]. Whereas a 50–60% increment in systemic insulin levels is associated with a 75–80% inhibition in splanchnic glucose output in non-obese controls, in obese subjects the fall in splanchnic glucose output is less than 40% (Figure 4.9). Furthermore, when the glucose infusion rate is increased in the obese subjects so as to cause an inhibition in splanchnic glucose output comparable to that observed in controls, it is achieved only in the face of a disproportionately greater rise in arterial insulin[32]. Thus as compared to normal weight subjects, in obesity comparable insulin increments are less effective in shutting off hepatic glucose output; comparable inhibition of hepatic glucose output requires an inordinate increase in serum insulin. It should be noted that these relationships between insulin levels and hepatic glucose balance are based upon systemic (arterial) insulin measurements rather than comparisons of portal venous levels. However, inasmuch as the liver is the major site of insulin degradation, and since turnover is if anything increased in obesity[72], it is likely that the portal-peripheral gradients observed in non-obese subjects[1] apply to the obese state as well.

With respect to the mechanism whereby obesity results in decreased tissue responsiveness to insulin, an increase in cell size has been implicated in the insulin resistance of adipose tissue[64]. In the case of the liver, an increase in the size of hepatocytes and augmented deposition of fat in the liver is not uniformly observed in obese subjects. On the other hand, studies in obese animals have demonstrated a decrease in the binding of insulin to its receptors on the liver cell membrane[73]. A similar decrease in insulin binding has been reported in mononuclear blood cells from obese humans[74]. It has recently been suggested that persistent hyperinsulinaemia (perhaps initially engendered by increased food intake) may of itself be responsible for decreased insulin binding by receptors[75]. Insulin resistance thus appears to be a fairly generalised phenomenon in obesity which is not restricted to the fat cell but involves the liver (and muscle tissue) as well, and may be a consequence of dietary-induced hyperinsulinaemia.

LIVER DISEASE

Considering the importance of the liver in glucose homeostasis, hepatocellular disease would be expected to alter blood glucose regulation. Interestingly, both hypoglycaemia and diabetes have been reported in association with disorders of the liver, the nature of the glucoregulatory defect depending on the type of liver disease. In general, hypoglycaemia has been noted in association with acute inflammatory processes of the liver (viral or toxic hepatitis), while glucose intolerance and hyperglycaemia are recognised complications of chronic liver failure (cirrhosis).

Viral hepatitis

The incidence of hypoglycaemia in viral hepatitis has been reported as ranging between 10 and 50%[76][77]. In a recent prospective analysis, over 50% of patients with viral hepatitis had fasting plasma glucose levels below 60 mg/100 ml[77]. None of the patients had evidence of massive or subacute hepatic necrosis either by biopsy or clinical criteria, indicating that hypoglycaemia can occur in the absence of severe hepatic damage.

116

However, it should be noted that in such patients, hypoglycaemia is generally mild (45–60 mg/100ml) and asymptomatic[77]. Severe symptomatic hypoglycaemia requiring treatment with glucose is generally seen in patients with massive hepatic necrosis[78][79], often as a terminal event.

With regard to the mechanism of hypoglycaemia in mild hepatitis, a fall in serum insulin appropriate to the low plasma glucose levels has been observed[77]. On the other hand, decreased hepatic glycogen stores, diminished responsiveness to glucagon, and failure of glycogen repletion following a high carbohydrate intake[77] suggest that hepatocellular damage results in abnormalities of glucoregulatory pathways within the liver. Interestingly such disturbances may occur despite clinical evidence of grossly intact protein synthetic mechanisms as indicated by normal levels of serum albumin and a normal prothrombin time[77].

In occasional patients with massive heptatic necrosis secondary to fulminant viral or toxic hepatitis, hyperinsulinaemia has been noted[79]. In such patients treatment with hypertonic glucose infusions in doses of 40 g or more per hour may be required to maintain a normal plasma glucose level. It should be recalled that glucose uptake by the brain and other non-insulin dependent tissues does not exceed 8–10 g per hour (Figure 4.1). It is thus clear that in contrast to the milder disturbances in blood glucose, hyperinsulinaemic hypoglycaemia is not solely due to failure of hepatic glucose production but reflects an increased rate of glucose utilisation as well. With respect to the mechanism of the hyperinsulinaemia, failure of the normal insulin-degradative function of the liver may be of some importance. On the other hand, increased secretion of insulin has not been excluded.

In addition to the reports of hypoglycaemia in hepatitis, hypoglycaemia has also been observed in occasional patients with cardiac failure and severe passive congestion of the liver[81][82]. In such circumstances diminished hepatic blood flow may be responsible for decreased delivery of gluconeogenic substrates and diminished release of glucose into the systemic circulation.

Cirrhosis

In patients with Laennec's cirrhosis, frank diabetes has been noted with an incidence of 15–30%[83-85]. The rate of occurrence of milder impairment in glucose tolerance is as high as 80%[86]. A variety of pathogenic factors have been implicated in 'hepatogenous diabetes', including hypokalaemia[87], elevated levels of growth hormone[88], and insulin resistance[89]. In the majority of such patients, serum insulin levels are elevated[90], suggesting

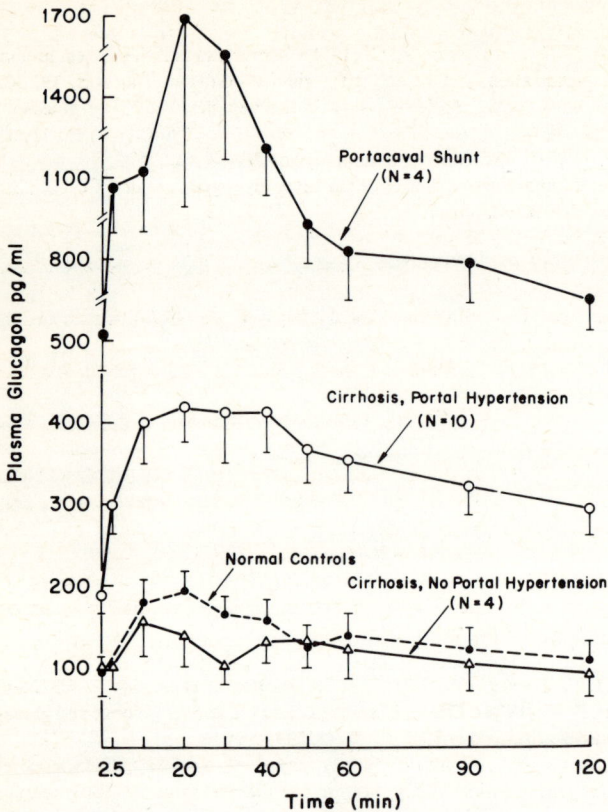

Figure 4.10 Plasma glucagon levels in normal subjects and patients with Laennec's cirrhosis. Glucagon concentrations were normal in cirrhotics without evidence of portal hypertension. In contrast, cirrhotics with spontaneous portal-systemic shunting (portal hypertension) were hyperglucagonemic. The most marked elevations in plasma glucagon were observed in patients with surgically induced, portacaval shunts. Based on the data of Sherwin *et al.*[92].

that decreased responsiveness to insulin rather than insulin deficiency is the primary factor responsible for glucose intolerance. With regard to the mechanism of insulin resistance in cirrhosis, recent studies have demonstrated hyperglucagonaemia in such patients[91], particularly in association with portal hypertension (Figure 4.10)[92]. It has been suggested that spontaneous or surgically-induced portal systemic shunting may lead to hypersecretion of glucagon as a consequence of accumulation of some as yet unidentified pancreatic α-cell secretogogue[92].

References

1. Blackard, W.G. and Nelson, N. C. (1970). Portal and peripheral vein immunoreactive insulin concentrations before and after glucose infusion. *Diabetes,* **19,** 302
2. Madison, L. L. (1969). Role of insulin in the hepatic handling of glucose. *Arch. Int. Med.,* **123,** 284
3. Felig, P. (1973). The Glucose-alanine cycle. *Metab.,* **22,** 179
4. Hultman, E., and Nilsson, L. H. (1971). Liver glycogen in man. Effect of different diets and muscular exercise. *Advan. Exp. Med. Biol.,* **11,** 143
5. Reichard, G. A., Jr., Moury, N. F., Hochella, N. J., Patterson, A. L. and Weinhouse, S. (1963). Quantitative estimation of the Cori cycle in the human. *J. Biol. Chem.,* **238,** 495
6. Felig, P., Owen, O. E. Wahren, J. and Cahill, G. F., Jr.(1969). Amino acid metabolism during prolonged starvation. *J. Clin. Invest.,* **48,** 584
7. Felig, P., Pozefsky, T., Marliss, E. and Cahill, G. F., Jr. (1970). Alanine: key role in gluconeogenesis. *Science,* **167,** 1003
8. Felig, P. and Wahren, J. (1971). Amino acid metabolism in exercising man. *J. Clin. Invest.,* **50,** 2703
9. Odessey, R., Khairallah, E. A. and Goldberg, A. L. (1974). Origin and possible significance of alanine production by skeletal muscle. *J. Biol. Chem.,* **249,** 7623
10. Mallette, L, E., Exton, J. H. and Park, G. R. (1960). Control of gluconeogenesis from amino acids in the perfused rat liver. *J. Biol. Chem.,* **244,** 5713
11. Cahill, G. F., Jr., Herrera, M. G., Morgan, A. P., Soeldner, J. S., Steinke, J., Levy, P. L., Reichard, G. A. , Jr. and Kipnis, D. M. (1966). Hormone-fuel interrelationships during fasting. *J. Clin. Invest.,* **45,** 1751

12. Gerich, J. E., Lorenzi, M., Schneider, V., Karam, J. H. Rivier, J., Guillemin, R. and Forsham, P. H. (1974). Effects of somatostatin on plasma glucose and glucagon levels in diabetes mellitus. *Nes Engl. J. Med.,* **291,** 544
12. Sakurai, H., Dobbs, R. and Unger, R. H. (1974). Somatostatin-induced changes in insulin and glucagon secretion in normal and diabetic dogs. *J. Clin. Invest.,* **54,** 1395
14. Felig, P., Marliss, E., Owen, O. E. and Cahill, G. F., Jr. (1969). Blood glucose and gluconeogenesis in fasting man. *Arch. Int. Med.,* **123,** 292
15. Marliss, E. B., Aoki, T. T., Unger, R. H., Soeldner, J. S. and Cahill, G. F., J. (1970). Glucagon levels and metabolic effects in fasting man. *J. Clin. Invest.,* **49,** 2256
16. Pozefsky, T., Tancredi, R. G., Moxley, R. T., Dupre, F. and Tobin, J. (1974), Forearm tissue metabolism in postabsorptive and 60 hour fasted man: studies with glucagon. *J. Clin. Invest.,* **53,** 61a (Abstract)
17. Owen, O. E., Felig, P., Morgan, A. P., Wahren, J. and Cahill, G. F., Jr. (1969). Liver and kidney metabolism during prolonged starvation. *J. Clin. Invest.,* **48,** 574
18. Benedict, F. G. (1915). *A Study of Prolonged Fasting,* Carnegne Institute of Washington, publication No. 203
19. Felig, P., Marliss, E. , Owen, O. E. and Cahill, G. F., Jr. (1969). Role of substrate in the regulation of hepatic gluconeogenesis. *Advan. Enzyme Reg.,* **7,** 41
20. Felig, P. (1972). Interaction of insulin and amino acids in the regulation of gluconeogenesis. *Israel J. Med. Sci.,* **8,** 262
21. Sherwin, R., Hendler, R. and Felig, P. (1975). Effect of ketone infusions on amino acid and nitrogen metabolism in man. *J. Clin. Invest.* **55,** 1382

22. Owen, O. E., Morgan, A. P., Kemp, H. G., Sullivan, J. M., Herrera, M. H. and Cahill, G. F. Jr. (1967). Brain metabolism during fasting. *J. Clin. Invest.*, **46,** 1589

23. Felig, P., Wahren, J. and Hendler, R. (1975). Influence of oral glucose ingestion on splanchnic glucose and gluconeogenic substrate metabolism in man. *Diabetes,* **24,** 468

24. Scow, R. L. and Cornfield, J. (1954). Quantitative relationships between the oral and intravenous glucose tolerance curves. *Amer. J. Physiol.,* **179,** 435

25. Perley, M. J. and Kipnis, D. M. (1967). Plasma insulin responses to oral and intravenous glucose: studies in normal and diabetic subjects. *J. Clin. Invest.,* **46,** 1954

26. Levine, R. and Fritz, I. B. (1956). The relation of insulin to liver metabolism. *Diabetes,* **5,** 209

27. Felig, P. and Wahren, J. (1971). Influence of endogenous insulin secretion on splanchnic glucose and amino acid metabolism in man. *J. Clin. Invest.,* **50,** 1702

28. Jackson, R. A., Peters, N., Advani, U., Perry, G., Rogers, J., Brough, W. and Pilkington, T. R. E. (1973). Forearm glucose uptake during the oral glucose tolerance test in normal subjects. *Diabetes,* **22,** 442

29. Pozefsky, T., Felig, P., Tobin, J. D., Soeldner, J. and Cahill, G. F., Jr. (1969). Amino acid balance across tissues of the forearm in postabsorptive man. Effects of insulin at two dose levels. *J. Clin. Invest.,* **48,** 2273

30. Felig, P., Marliss, E. and Cahill, G. F., Jr. (1969). Plasma amino acid levels and insulin secretion in obesity. *New Engl. J. Med.,* **281,** 811

31. Bagdade, J. D., Bierman, E. L. and Porte, D., Jr. (1967). The significance of basal insulin levels in the evaluation of the insulin response to glucose in diabetic and non-diabetic subjects. *J. Clin. Invest.,* **46,** 1549

32. Felig, P., Wahren, J., Hendler, R. and Brundin, T. (1974). Splanchnic glucose and amino acid metabolism in obesity. *J. Clin. Invest.,* **53,** 582

33. Liljenquist. J. E. Chiasson, J. L., Finger, R. E. and Lacy, W. W. (1974). Differential effect of insulin on glycogenolysis and gluconeogenesis. *Diabetes,* **23,** 349 (Abstract)

34. Andres, R., Cader, G. and Zierler, K. L. (1956). The quantitatively minor role of carbohydrate in oxidative metabolism by skeletal muscle in intact man in the basal state. Measurements of oxygen and glucose uptake and carbon dioxide and lactate production in the forearm. *J. Clin. Invest.,* **35,** 671

35. Rabinowitz, D. and Zierler, K. L. (1962). Role of free fatty acids in forearm metabolism in man quantitated by use of insulin. *J. Clin Invest.,* **41,** 2191

36. Wahren, J., Felig, P., Ahlobrog, G. and Jorfeldt, L. (1971). Glucose metabolism during leg exercise in man. *J. Clin. Invest.,* **50,** 2715

37. Rowell, L. B., Masoro, E. J. and Spencer, M. J. (1965). Splanchnic metabolism in exercising man. *J. Appl. Physiol.,* **20,** 1032

38. Bergstrom, J. and Hultman, E. (1967). A study of glycogen metabolism during exercise in man. *Scand. J. Clin. Lab. Invest.,* **19,** 218

39. Ahlborg, G., Felig, P., Hagenfeldt, L., Hendler, R. and Wahren, J. (1974). Substrate turnover during prolonged exercise in man. Splanchnic and leg metabolism of glucose, free fatty acids and amino acids. *J. Clin. Invest., 53,* 1080

40. Young, D. R., Pelligra, R., Shapira, J., Adachi, R. R. and Skrettingland. (1967). Glucose oxidation and replacement during prolonged exercise in man. *J. Appl. Physiol.,* **23,** 734

41. Felig, P., Wahren, J., Hendler, R. and Ahlborg, G. (1972). Plasma glucagon levels in exercising man. *New Engl. J. Med.,* **287,** 184

42. Felig, P., Wahren, J. and Hendler, R. (1974). Senstivity of hepatic gluco-regulatory

mechanisms in exercising man. *Proc, 56th Annual Meeting of the Endocrine Society, Atlanta.*

43. Malins, J. M. (1963). Glucose tolerance and glycosuria in the general population. *Brit. Med. J.,* **2,** 655

44. Cerasi, E. and Luft, R. (1967). Insulin response to glucose infusion in diabetic and non-diabetic monozygotic twin pairs. *Acta Endocrinol.,* **55,** 330

45. Seltzer, H. S., Allen, E. W., Herron, A. L., Jr. and Brennan, M. T. (1967). Insulin secretion in response to glycemic stimulus: relation of delayed initial release to carbohydrate intolerance in mild diabetes mellitus. *J. Clin. Invest.,* **46,** 323

46. Muller, W. A., Faloona, G. R., Aguilar-Parada, E. and Unger, R. H. (1970). Abnormal alpha cell function in diabetes. Response to carbohydrate and protein ingestion. *New Engl. J. Med.,* **282,** 109

47. Myers, J. D. (1950). Net splanchnic glucose production in normal man and in various disease states. *J. Clin Invest.,* **29,** 1421

48. Bearn, A. G., Billing, B. H. and Sherlock, S. (1951). Hepatic glucose output and hepatic insulin sensitivity in diabetes mellitus. *Lancet,* **ii,** 698

49. Wahren, J., Felig, P., Cerasi, E. and Luft, R. (1972). Splanchnic and peripheral glucose and aminoacid metabolism in diabetes. *J. Clin. Invest.,* **51,** 1870

50. Shreeve, W. W., Baker, N., Miller, M., Shipley, R. A., Incefy, G. E. and Craig, J. W. (1956). ^{14}C studies in carbohydrates metabolism. II. The oxidation of glucose in diabetic human subjects. *Metab. Clin. Exp.,* **5,** 22

51. Reichard, G. A., Jr., Moury, N. F., Hochella, N. J., Patterson, A. L. and Weinhouse, S. (1963). Quantitative estimation of the Cori cycle in the human. *J. Biol. Chem.,* **238,** 495

52. Manougian, E., Pollycove, M., Linfoot, J. A. and Lawrence, J. H. (1964). ^{14}C glucose kinetic studies in normal, diabetic and acromegalic subjects. *J. Nucl. Med.,* **5,** 763

53. Forbath, N. and Hetenyi, G. Jr. (1966). Glucose dynamics in normal subjects and diabetic patients before and after a glucose load. *Diabetes,* **15,** 778

54. Bodny, P. K., Bloom, W. L., Whitner, V. S. and Farrar, B. W. (1949). Studies of the role of the liver in human carbohydrate metabolism by the venous catheter technic. II. Patients with diabetic ketosis, before and after the administration of insulin. *J.Clin. Invest.,* **28,** 1126

55. Felig, P., Marliss, E., Ohman, J. L. and Cahill, G. F., Jr. (1970). Plasma amino acid levels in diabetic ketoacidosis. *Diabetes,* **19,** 727

56. Müller, W. A., Faloona, G. R. and Unger, R. H. (1973). Hyperglucagonemia in diabetic ketoacidosis. Its prevalence and significance. *Amer. J. Med.,* **54,** 52

57. McGarry, J. D. and Foster, D. W. (1972). Regulation of ketogenesis and clinical aspects of the ketotic state. *Metab.,* **21,** 471

58. McGarry, J. D. and Foster, D. W. (1973). Acute reversal of experimental diabetic ketoacidosis in the rat with (+)–decanoylcarnitine. *J. Clin. Invest.,* **52,** 877

59. Sanders C. A., Levinson, G. E., Abelmann, W. H. and Freinkel, N. (1964). Effect of exercise on the peripheral utilisation of glucose in man. *New Engl. J. Med.,* **271,** 220

60. Wahren, J., Hagenfeldt, L. and Felig, P. (1975). Splanchnic and leg exchange of glucose, amino acids and free fatty acids during exercise in diabetes mellitus. *J. Clin. Invest.* **55,** 1303

61. Kemp, C. B., Knight, M. J., Scharp, D. W., Laly, P. E. and Ballinger, W. F. (1973). Transplantation of isolated pancreatic islets into the portal vein of diabetic rats. *Nature,* **244,** 447

Diabetes

62. Albisser, A. M., Leibel, B. S., Ewart, T. G., Davidovac, Z., Botz, C.K. and Zingg' W. (1974). An artificial endocrine pancreas. *Diabetes,* **23**, 389
63. Felig, P. (1974). Insulin: rates and routes of delivery. *New Engl. J. Med.,* **291**, 1031
64. Salans, L. B., Knittle, J. L. and Hirsch, J. (1968). The role of adipose cell size and adipose tissue insulin sensitivity in the carbohydrate intolerance of human obesity. *J. Clin. Invest.,* **47**, 153
65. Rabinowitz, D, and Zierler, K. L. (1962). Forearm metabolism in obesity and its response to intra-arterial insulin: characterization of insulin resistance and evidence for adaptive hyperinsulinism. *J. Clin. Invest.,* **41**, 2173
66. Ogilvie, R. F. (1935). Sugar toleerance in obese subjects: review of 65 cases. *Quart. J. Med.,* **4**, 345
68. Kreisberg. R. A. (1968). Glucose metabolism in normal and obese subjects. Effect of phenformin. *Diabetes,* **17**, 481
68. Kreisberg, R. A. (1968). Kintetics of glucose utilization in obesity: the effect of phenformin. *Ann. N. Y. Acad. Sci.,* **148**, 743
69. Shigeta, Y., Oji, N., Hoshi, M. and.King, M. (1965). Fatty acid synthesis and gluconeogenesis in rats fed a high caloric diet. *Metab. Clin. Exp.,* **15**, 761
70. Shreeve, W. W., Hoshi, M., Ohi, N., Shigeta, Y. and Abe, H. (1968). Insulin and the utilization of carbohydrate in obesity. *Amer. J. Clin. Nutr.,* **21**, 1404
71. Bortz, W. M., Paul P., Haff, A. C. and Holmes, W. I. (1972). Glycerol turnover and oxidation in man. *J. Clin. Invest.,* **51**, 1537
72. Genuth, S. M. (1972). Metabolic clearance of insulin in man. *Diabetes,* **21**, 1003
73. Kahn, C R., Neville, D. M., Jr. and Roth, J. (1973). Insulin-receptor interaction in the obese-hyperglycemic mouse. *J. Biol. Chem.,* **248**, 244
74. Archer, J. A., Gorden, P. and Roth, J. (1975). Defect in insulin binding to receptors in obese man. Amelioration with caloric restriction. *J. Clin. Invest.,***55**, 166
75. Kahn, C. R., Soll, A., Neville, D. M., Jr. and Roth, J. (1973). Severe deficiency in insulin receptors: a common denominator in the insulin resistance of obesity. *Clin. Res.,* **21**, 628 (Abstract)
76. Zimmerman, H. J., Thomas, L. J. and Scherr, E. H. (1953). Fasting blood sugar in hepatic disease with reference to infrequency of hypoglycemia. *Arch. Intern. Med.,* **91**, 577
77. Felig, P., Brown, W. V., Levine, R. A. and Klatskin, G. (1970). Glucose homeostasis in viral hepatitis. *New. Engl. J. Med.,* **283**, 1436
78. Marks, V. and Rose, F. C. (1965). *Hypoglycaemia.* pp. 166-172 (Oxford: Blackwell Scientific Publications)
79. Samson, R. I., Trey, C., Timme, A. H. and Saunders, S. (1967). Fulminating hepatitis with recurrent hypoglycemia and hemorrhage. *Gastroenterology,* **53**, 291
80. Lucke. B. and Mallory, T. (1946). The fulminant form of epidemic hepatitis. *Amer. J. Pathol.,* **22**, 867
81. Melinkoff, S. M. and Tumulty, P. A. (1952). Hepatic hypoglycemia. Its occurrence in congestive heart failure. *New Engl. J. Med.,* **247**, 745
82. Blcok, M. B., Gambetta, M., Resnkov, L. and Rubinstein, A. H. (1972). Spontaneous hypoglycaemia in congestive heart failure. *Lancet,* **ii,** 736
83. Hed, R. (1958). Clinical studies in alcoholism. I Incidence of diabetes mellitus in portal cirrhosis. *Acta Med. Scand.,* **162**, 189
84. Megyesi, C., Samols, E. and Marks, V. (1967). Glucose intolerance and diabetes in chronic liver disease. *Lancet.* **ii,** 1051

85. Conn, H. O., Schreiber, W., Elkington, S. G. and Johnson, T. R. (1969). Cirrhosis and diabetes. I. Increased incidence of diabetes in patients with Laennec's cirrhosis. *Amer. J. Digestive Dis.*, **14**, 837

86. Conn, H. O., Schreiber, W. and Elkington, S. G. (1971). Cirrhosis and diabetes. II. Association of impaired glucose tolerance with portal-systemic shunting in Laennec's cirrhosis. *Amer. J. Digestive Dis.*, **16**, 227

87. Podolsky, S., Zimmerman, H. J. and Burrows, B. A. (1973). Potassium depletion in hepatic cirrhosis: a reversible cause of impaired growth-hormone and insulin response to stimulation. *New. Engl. J. Med.*, **288**, 644

88. Conn, H. O. and Daughaday, W. H. (1970). Cirrhosis and diabetes. V. Serum human growth hormone levels in cirrhosis. *J. Lab. Clin. Med.*, **76**, 678

89. Felig, P. and Wahren, J. (1975). Etiology of glucose intolerance in liver disease: an hypothesis based upon observations in normal, diabetic and prediabetic subjects. *Proceedings, Second International Beilinson Symposium on Diabetes in Juveniles* (Edited by Z. Laron and M. Karp) (in press)

90. Collins, J. R. and Crofford, O. B. (1969). Glucose intolerance and insulin resistance in patients with liver disease. *Arch. Intern. Med.*, **12**, 142

91. Marco, J., Diego, J. and Villanueva, M. L. (1973). Elevated plasma glucagon levels in cirrhosis of the liver. *New Engl. J. Med.*, **289**, 1107

92. Sherwin, R., Joshi, P., Hendler, R., Felig, P. and Conn, H. O. (1974). Hyperglucagonemia in Laennec's cirrhosis. The role of portal-systemic shunting. *New Engl. J. Med.*, **290**, 239

ACKNOWLEDGEMENTS
Dr Felig is recipient of a Research Career Development Award (AM 70219) from the National Institutes of Health, U.S. Public Health Service. The research conducted in the author's laboratory referred to in this chapter was supported by grants AM 13526 and RR 125 from the National Institutes of Health.

5
ATHEROSCLEROSIS AND DISORDERS OF LIPID METABOLISM IN DIABETES

R. W. Stout, E. L. Bierman and J. D. Brunzell

In the half century since insulin became available for the treatment of diabetes there has been a remarkable change in the course of this disease. The average life span of diabetics has been greatly increased[1] and death from uncontrolled hyperglycaemia and ketosis has become rare. However, the diabetic still has a considerably shorter life expectancy than the non-diabetic[2], and a higher mortality at all ages (Table 5.1). There has been a marked change in the causes of death in diabetics since insulin became available (Table 5.2), the commonest cause of death now being atherosclerotic vascular disease and its complications. This is also the commonest cause of death in the Western population as a whole[3].

Diabetes is associated with disease of both small and large vessels. The microangiopathy is a disorder of the capillaries and is characterised by thickening of the basement membrane (basal lamina)[4]. Although controversy exists over a possible relationship to the abnormal carbohydrate metabolism of diabetes, the lesions resulting from the thickened basal lamina of capillaries (retinal microaneurysms and nodular glomerularsclerosis) appear to be specific manifestations of diabetes. The large vessel complications of diabetes consist of atherosclerosis of the aorta and the larger muscular arteries, particularly the coronary and cerebral arteries, and the arteries supplying the lower limbs. Atherosclerosis in diabetics differs neither in distribution nor in morphological appearance from atherosclerosis in non-diabetics[5] and appears to be unrelated to the capillary disease. The large vessel complications of diabetes will be discussed in this chapter.

Table 5.1 Mortality of diabetic patients compared to general population (1950–1958[1]). Ratio of diabetics to general population death rate

Age (years)	Male	Female
5–14	1.8	2.7
15–24	4.5	1.3
25–34	7.4	13.9
35–44	4.4	4.5
45–54	2.1	3.1
55–64	1.8	2.4
65–74	1.6	2.4

Table 5.2 Causes of death in diabetics since 1922[1]. % total deaths

Cause of death	1922-29	1930-36	1937-43	1944-49	1950-55	1956-62
Diabetic coma	14.5	5.0	2.8	1.7	1.1	1.0
Total vascular	46.8	58.3	65.8	71.3	76.9	76.6
Cardiac	23.1	33.4	41.1	47.1	50.2	51.2
Cerebral	8.1	10.1	11.8	12.8	13.3	12.6
Gangrene	8.6	7.6	5.3	2.9	2.0	1.8
Infections	21.7	15.7	12.6	7.6	5.9	5.7
Cancer	7.4	9.4	9.0	9.7	10.1	10.5

A major component of the atherosclerotic lesion is lipid and much attention has been directed to abnormalities in the metabolism of circulating lipids in subjects with atherosclerotic vascular disease, in the hope that this will lead to a greater understanding of the pathogenesis of atherosclerosis. Abnormalities in circulating lipids are common in diabetics and are discussed in the second part of this chapter.

Diabetes has been defined as a disorder in which the level of blood glucose is persistently raised above the normal range[6]. However, in the general population, blood glucose values have a continuous frequency distribution and hence the range of normality has to be arbitrarily defined. A variety of criteria for the diagnosis of diabetes has been described[7] based on the results of glucose tolerance tests. Within the syndrome of diabetes thus defined, there are many variations in clinical presentation and biochemical abnormality. It is acknowledged that genetic diabetes mellitus is presumably present from birth, but is difficult to recognise prior to the onset of hyperglycaemia. For the purposes of the discussion on atherosclerosis, diabetes, hyperglycaemia and glucose intolerance are regarded as synonymous. However, in the sections on lipid disorders, abnormalities in lipid metabolism in glucose intolerance and in overt diabetes with fasting hyperglycaemia will be discussed separately.

ATHEROSCLEROSIS AND DIABETES

In the last fifty years, the frequency of atherosclerosis in diabetics has been the subject of many autopsy [8-14] or clinical [15-23] surveys. Almost all report that ischaemic heart disease is considerably more prevalent in diabetics than non-diabetics. In particular, the prevalence of atherosclerosis in diabetic women is the same as that in diabetic men at all ages, in contrast to the experience in non-diabetics, where atherosclerosis is much less frequent in young women[24]. Some reports suggest that vascular disease occurs at a younger age in diabetics than non-diabetics, but this has not been confirmed[25]. In addition diabetics and subjects with glucose intolerance have a considerably higher mortality from ischaemic heart disease than those with normal glucose tolerance [26 27].

The concept that atherosclerosis is more common in diabetics than non-diabetics has recently been challenged by the results of two careful studies. In the Oxford necropsy study of arterial disease [28], the area of aorta involved by atherosclerosis was measured in a large number of autopsy specimens. There was only a small number of diabetics in the series, and only enough female diabetics for a numerical comparison to be made with non-diabetics. It was found that whereas cardiac infarction was very common in diabetic women, diabetics had the same degree of aortic atherosclerosis as non-diabetics when the groups were matched for the presence or absence of cardiac infarction. This suggested that diabetics differ in their susceptibility to thrombotic arterial occlusion rather than in disease of the arterial wall. On the clinical side, careful assessment of the vascular status of patients attending the large diabetic clinic at King's College Hospital, London [25] did not reveal a high incidence of coronary disease in young diabetics. However, the equal frequency of atherosclerosis in male and female diabeticc at all ages was confirmed.

Attention also has been directed to carbohydrate metabolism in patients with ischaemic vascular disease in the absence of overt or previously diagnosed diabetes. A recent review has reported that about 61% of 590 oral glucose tolerance tests and 55% of 557 intravenous glucose tolerance tests were abnormal in patients with atherosclerotic vascular disease [29].

It may be argued that glucose intolerance in ischaemic heart disease is a non-specific reaction to the stress of an acute illness. However, repeat glucose tolerance testing weeks[30] or months[26] after the acute episode reveals essentially the same results as tests performed soon after the illness. Recently the problem of diagnosing coronary atherosclerosis accurately during life has been tackled by the use of coronary arteriography; in this way, arterial disease can be demonstrated before ischaemic changes occur in the myocardium. In each of two studies [31] [32], 66% of subjects with demonstrable coronary atheroma but without myocardial infarction had abnormal glucose tolerance compared with 18–25% in those with normal arteries. These lines of evidence suggest that glucose intolerance after myocardial infarction is not related to the stress of an acute ischaemic episode.

There are risks inherent in drawing conclusions from autopsy series concerning the chance of two diseases co-existing, particularly if the association of the two diseases in the same person increases the chance of hospital admission or death [33]. Interpretation of clinical studies of the relationship between atherosclerosis and other diseases is also difficult [34]. Thus the problem of making a clinical diagnosis of atherosclerosis in the absence of its complications, combined with the near universal presence of the disease in man makes group comparison difficult as there are virtually no normal controls. As both diabetes and advanced atherosclerosis tend to bring the subject to medical attention, the possibilty of the other condition being recognised is increased. Studies carried out in large unselected populations are therefore particularly interesting and important.

Table 5.3a Prevalence of hyperglycaemia among persons with vascular disease in Tecumseh, Michigan[36]

	Male *o / e*	Female *o / e*
Coronary heart disease	1.67	1.82
Peripheral and cerebral vascular disease	1.49	1.90
T-wave changes	1.77	1.92

Table 5.3b Vascular disease in diabetics in Tecumseh, Michigan[36]

Age	Male *o / e*	Female *o / e*
20–39	0	0
40–59	3.33	2.73
60+	1.57	1.73
Total	1.90	1.84

In the Framingham study [35] it was found that both the morbidity and mortality from cardiovascular disease was significantly higher in diabetics than in non-diabetics and was excessive in all diabetic subgroups studied. Although some of the factors associated with cardiovascular disease, including hypertension, obesity and hypercholesterolaemia were present in the diabetics, it was concluded that they could not completely explain the high incidence of the disease in diabetics. In the Tecumseh prevalence survey, the proportion of subjects with elevated blood sugar levels one hour after ingestion of glucose was significantly higher for the subjects with cardiovascular disease than those without [36] (Table 5.3a). Conversely, the diabetics identified in the survey had a higher prevalence of vascular disease than non-diabetics of the same age and sex (Table 5.3b). It was concluded that hyperglycaemia is an independent risk factor for atherosclerosis and was at least as important as hypertension, hypercholesterolaemia or excessive body weight. Preliminary results for the incidence of ischaemic heart disease in hyperglycaemic subjects in Tecumseh confirm the results of the prevalence studies [37]. In Bedford [38], participants with glycosuria were classified as diabetic, borderline or normal according to their blood sugar level two hours after 50 g of glucose orally. The age-adjusted prevalence of both symptoms and electrocardiographic changes of ischaemic heart disease was lowest in the control group, intermediate in the borderline group and highest in the diabetics. Follow-up studies on the same subjects show that the incidence of cardiovascular disease is similarly related to the initial blood sugar level [39]. A recent survey in an industrial population also revealed a higher incidence of mortality from cardiovascular disease in diabetics than in matched controls [40]. In Seattle a high prevalence of diabetes was found in unselected three-month myocardial infarction survivors in a total community [41]. The results of these large epidemiological studies are supported by the results of the International Atherosclerosis Project [42] in which a large number of arteries from several countries were examined by standardised techniques. The coronary arteries and abdominal aortae of diabetics showed more atherosclerosis than those of non-diabetics, regardless of sex, age, race or geographic location.

There is thus a large body of evidence linking hyperglycaemia and diabetes with atherosclerotic vascular disease. Although there are difficulties in the interpretation of some of the data and some of it is contradictory, the recent epidemiological studies indicate that diabetics have an increased susceptibility to atherosclerosis. Female diabetics seem to have a particularly high risk as, in contrast to non-diabetics, the prevalence of atherosclerosis in diabetics is equal in both sexes at all ages.

If hyperglycaemia and atherosclerosis are associated, it may be possible to identify other abnormalities which are common to both diseases and which may be involved in the pathogenesis of the vascular disease. Attention will now be paid to the characteristics of the diabetic state itself and to some of the 'risk factors' for atherosclerosis that are commonly found in diabetics. The Task Force on Arteriosclerosis of the National Heart and Lung Institute [3] identified eight risk factors, including diabetes. Of these, there is no information on the role of smoking, physical inactivity, genetic aspects or psychosocial stress on the association of diabetes and atherosclerosis. Abnormalities of lipid metabolism, hypertension and obesity have been studied in diabetics and will be discussed, as will characteristics associated with diabetes itself.

Factors which may influence atherosclerosis in diabetes

(a) Duration, severity and treatment of diabetes

Diabetes is a disorder which can only be diagnosed on the basis of investigations performed by a doctor and the diagnosis of diabetes leads almost invariably to the institution of some form of treatment. It is thus difficult to differentiate the influence of the disease from the influence of treatment on the development of complications. In many cases, treatment of diabetes allows the patient to survive long enough to develop the chronic complications. Furthermore, in most cases, it is impossible to determine the date of onset of diabetes, particularly as glucose intolerance may be a late manifestation of a complex disorder, which may have been present from birth. Frequently the presence of diabetes is discovered when the patient seeks medical assistance for some other condition, and it is clear from the many reports of a high frequency of previously undiagnosed diabetes and glucose intolerance in patients presenting with vascular disease, that atherosclerosis in many diabetics is unrelated to antidiabetic therapy.

Atherosclerosis is a slowly progressive disorder, becoming clinically manifest with advancing years, and it would be expected that there would be a relation between the duration of diabetes and the incidence of atherosclerosis. Such a relationship has been described by some [11][14][43][44] but not all authors [13][25][45-47]. As the incidence of both glucose intolerance and arterial disease increases with advancing years, it may be age rather than duration of diabetes that is important [13].

There appears to be little [14] or no [11][13][45] relationship between the frequency of the large vessel complications and the severity of diabetes, as measured by the degree of hyperglycaemia or the treatment required. Nor is the degree of 'control' of diabetes by treatment related to the occurrence of atherosclerotic heart disease [13]. Furthermore, the large prospective study organised by the University Group Diabetes Program [48] on the treatment of diabetes showed that reduction of blood sugar by any means was not effective in preventing the onset of the atherosclerotic vascular complications of diabetes. Indeed two groups of subjects, those taking tolbutamide or phenformin had a higher mortality from cardiovascular disease than those taking either insulin or a placebo. Various aspects of these trials have been criticised [49], and smaller trials with tolbutamide in diabetics [50][51] and with phenformin in non-diabetics [52] have not confirmed the deleterious effects of these drugs. Nevertheless, it seems clear from the results of the UGDP study that reduction of blood sugar does not prevent the cardiovascular complications of diabetes. Of the many explanations offered for these findings, it seems likely that either the vascular complications of diabetes are related to some aspect of the disease which is not influenced by currently available treatment, or that the vascular disease had already reached an irreversible stage before treatment was instituted.

The fact that vascular disease in diabetes appears to have little relationship to the severity or duration of the disease and is not benefited by current therapeutic regimes, suggests that it is unlikely that hyperglycaemia itself has an important role in the pathogenesis of atherosclerosis in diabetics.

(b) Abnormal lipid metabolism
Abnormalities of lipid metabolism in diabetes are fully discussed in another section of this chapter. However, a brief discussion of the relationship of hyperlipidaemia to atherosclerosis in diabetes is relevant here.

Elevations of triglyceride and cholesterol levels are independently related to atherosclerotic vascular disease in the population as a whole [53]. However, triglyceride abnormalities appear to be considerably more common than hypercholesterolaemia in diabetics with arterial disease [13 46 47 54]. In a recent study it was confirmed that hypertriglyceridaemia was much commoner than hypercholesterolaemia in diabetics with atherosclerosis; however diabetics without detectable atherosclerosis had triglyceride and cholesterol levels indistinguishable from non-diabetic non-atherosclerotic subjects [55] (Table 5.4). Although triglyceride levels correlated with body weight, it was concluded that obesity was not the sole factor determining triglyceride levels and that triglyceride levels and obesity were independently related to atherosclerosis in diabetes. In the Framingham study [35] serum cholesterol levels were no higher in diabetic men, but were slightly higher in diabetic women than in the total study group. The high incidence of cardiovascular disease in diabetics could not be explained by changes in cholesterol levels. In Tecumseh [56] a group of subjects who had both cardiovascular disease and hyperglycaemia initially, had significantly higher triglyceride levels than subjects with only one or neither of these abnormalities. Serum cholesterol and blood pressure were similar in all groups.

There have been no prospective studies on the predictive value of hypertriglyceridaemia in diabetic vascular disease. However, serum cholesterol elevation has been reported to have some predictive value for atherosclerosis in diabetics in a small number of subjects studied over a five-year period [57].

Thus, abnormalities in serum lipids appear to be related to atherosclerotic vascular disease in diabetics in an analogous fashion to non-diabetics, but abnormalities of neither triglyceride nor cholesterol regulation

Table 5.4 Lipid levels in diabetics (age 30–59 years) in relation to atherosclerosis[55]

	Control	Diabetic without atherosclerosis	Diabetic with atherosclerosis
Triglyceride >150 mg/100 ml	19%	15%	50%
Mean triglyceride mg/100 ml	106	103	165
Cholesterol >250 mg/100 ml	10%	8%	23%
Mean cholesterol mg/100 ml	202	205	229

can adequately explain the increased frequency of atherosclerosis in diabetes.

(c)   Obesity
In general, morbidity and mortality are higher among the overweight than those whose weight is normal[58]. Despite this, some of the major epidemiological studies of coronary heart disease have not demonstrated an independent relationship between this condition and anything less than gross obesity[59]. However, obesity is a disorder associated with many abnormalities such as hypertriglyceridaemia[60], hypercholesterolaemia[61] and hypertension[62], which are themselves associated with vascular disease. The relationship between obesity and atherosclerosis is thus multifaceted and incompletely understood.

Obesity is considerably more common in diabetics than in the population as a whole[63] and vascular disease is more common in diabetics who have gained weight than in those who have not[47]. Diabetics who have gained weight but lost it again have a lesser frequency of vascular disease, less in those who have lost more weight than those who lost little weight[47]. Although another study of 383 diabetics revealed no relationship between atherosclerosis and obesity[11], Santen *et al.*[55] found more obese subjects in their diabetics with arterial disease than in those without the diabetic complications.

Obesity is closely related to circulating triglyceride levels in both diabetics [55] and non-diabetics [60], but triglyceride levels and obesity seem to be independently related to atherosclerosis in diabetics [55]. The relationship between obesity and triglyceride levels may be mediated by way of the elevated insulin levels associated with obesity [60]. There is a relationship between body weight and cholesterol levels[61] and obesity is associated with increased total body cholesterol synthesis [64] [65]. Hypertension is also related to body weight [62] [66] and blood pressure is reduced by weight reduction [66].

The role of obesity in the vascular complications of diabetes is not clear and the relationship may be indirect. While weight reduction results in improved glucose tolerance, lower triglyceride levels and often lower blood pressure, it is not known if the development of atherosclerosis is slowed or reversed by this measure.

(d) Hypertension

Hypertension has been shown to be an important risk factor for death from cardiovascular disease [3]. The risk progressively increases with increasing blood pressure and is diminished by therapeutic reduction of blood pressure [67] [68]. However, the disorders most closely connected with hypertension are cerebrovascular disease and cardiac failure [69], the relationship between hypertension and coronary atherosclerosis being less marked.

There is a widespread clinical impression that arterial pressure tends to be higher in diabetics than non-diabetics. However, apart from the relatively few diabetics with renal disease, the evidence linking diabetes with hypertension is contradictory. A large autopsy study [70] showed that hypertension, diagnosed as cardiac enlargement without valvular disease, was twice as common in diabetics than non-diabetics, and a survey of non-hospitalised subjects[71] found that hypertension was 54% more common in diabetics than non-diabetics matched for age, sex and social class. This study suggested that the increased susceptibility of diabetics to atherosclerotic heart disease could almost entirely be explained by a greater prevalence of hypertension. In the large population study in Tecumseh, Michigan [36], it was found that hypertension was more common than expected in diabetic women, but did not have a significantly increased frequency in diabetic men (Table 5.5a). Conversely, hyperglycaemia was more common in subjects with hypertension than in the population as a whole (Table 5.5b). However, the risk of hyperglycaemia for cardiovascular disease was independent of its association with hypertension, and was at least as important as hypertension [72]. Similar conclusions were drawn from the population study in Bedford, England [38], where it was found that only part of the association between hyperglycaemia and arterial disease could be explained by differing levels of blood pressure. In the Framingham study [35], blood pressure was significantly higher in diabetics in the study population as a whole, but the high incidence of cardiovascular disease in diabetics could not be entirely explained by this.

Table 5.5a Frequency of hypertension in diabetics: Tecumseh[36]

| Age | Male | | Female | |
	SHT o/e	DHT o/e	SHT o/e	DHT o/e
20–39	1.67	1.67	2.00	3.33
40–59	1.85	0.74	2.35	2.50
60+	0.94	1.18	1.45	1.19
Total	1.38	1.04	1.79	1.76

Table 5.5b Prevalence of hyperglycaemia among persons with hypertension: Tecumseh[36]

	Male o/e	Female o/e
Systolic hypertension	1.25	1.41
Diastolic hypertension	1.20	1.32

An excess prevalence of hypertension has not been found in all studies of diabetics [73]. A careful study of 1100 diabetics [74] showed no difference in the frequency of hypertension from age and sex matched controls, except at age 70–79 years, and suggested that the earlier clinical impression resulted from the high predominance of older diabetics. Another study showed a rise in blood pressure with increasing duration of diabetes [75].

Thus it seems that hypertension is only slightly more common in diabetics than in the general population and that age, sex, obesity, renal disease and duration of diabetes may influence the association of diabetes and hypertension. Although the co-existing presence of hypertension will increase the risk of atherosclerosis, the balance of evidence seems to be that hypertension alone cannot account for the increased experience of cardiovascular disease in diabetics.

(e) Diet

A retrospective study of diabetics managed at time periods two decades apart suggested that in the later period, an increase in the frequency of atherosclerotic complications was associated with an increase in the carbohydrate content of the diet [76]. Serum triglyceride levels were also higher in the later period and it was suggested that the atherosclerosis was related to carbohydrate-induced increases in triglyceride levels. However, more recent studies have failed to find a relationship between dietary carbohydrate and triglyceride levels in diabetics [55], despite a close association between hypertriglyceridaemia and atherosclerosis in these subjects. Furthermore, although increasing the proportion of carbohydrate in the diet induces acute elevation of circulating fasting triglyceride levels [77] this phenomenon appears to be transient in both diabetic [78] [79] and non-diabetic subjects [80], and restricted to the basal (overnight fasted) state, since triglyceride levels over a twenty-four hour period are actually lower in subjects consuming high carbohydrate diets [81]. Increasing dietary carbohydrate tends to improve glucose tolerance in mild diabetics [82], and lowers fasting glucose levels with no increase in glycosuria or the insulin requirements of severe diabetics [83]. In Japan, the average diet contains a higher proportion of carbohydrate than in Western countries, yet Japanese diabetics have very little coronary or peripheral atherosclerosis [39].

Diets which are low in carbohydrate must have a high content of fat if the total calories are to remain unchanged. Thus, the restricted carbohydrate diet, to which diabetics are frequently advised to adhere, has an excessively high fat content. Dietary fat and cholesterol have been considered to be related to hypercholesterolaemia and hence to atherosclerosis and it has been suggested that similar dietary factors may be responsible for the increased prevalence of atherosclerosis in diabetics. A prospective study on the effect of a low fat, high carbohydrate diet on the vascular complications of diabetes started about ten years ago [79] and the results are still awaited. However, the initial studies showed that the diabetics consuming a diet high in carbohydrate and low in fat and cholesterol had lower cholesterol levels and similar triglyceride levels to diabetics taking a standard 'diabetic diet'. The insulin requirements of the diabetics did not change when the carbohydrate content of the diet was increased.

The role of diet in the pathogenesis of atherosclerosis in diabetics remains unclear. However, there is little evidence to support the idea that a high carbohydrate diet is harmful and it seems likely that diet has the same role in atherosclerosis in diabetics as in non-diabetics.

(f) Insulin secretion

Glucose intolerance, the characteristic feature of diabetes, is associated with diminished insulin secretion in response to glucose stimulation [84]. This is seen in all diabetics if the effect of obesity on fasting or basal insulin secretion is eliminated by expressing the insulin response to glucose in terms of the basal insulin level [85]. However, the relationship between basal insulin secretion and adiposity is unaltered by the presence of diabetes. Thus obese diabetics usually have higher basal and stimulated insulin concentrations than thin non-diabetics, even though the insulin secretion in the diabetics is relatively impaired. Elevated circulating insulin levels have also been found in early [86] and mild [87] diabetes.

Diabetics who have no endogenous insulin response to glucose stimulation and thus have severe hyperglycaemia and a tendency to ketosis, are usually treated with exogenous insulin. There have been few measurements of circulating insulin levels in insulin-treated diabetics, chiefly because of technical difficulties caused by the patients' anti-insulin antibodies interfering with the immunoassay of insulin. The studies that have been performed have produced contradictory results. Some studies have shown fasting insulin levels that are normal or elevated, sometimes grossly, in insulin-treated diabetics 14–24 h after their last insulin injection[88-90]. Other workers reported abnormally low fasting free insulin levels in insulin-treated diabetics [91]. Clearly the technical problems in the measurements have to be resolved before firm conclusions can be drawn on the insulin levels in insulin-treated diabetics. Nor is it clear how much immunoassayable insulin is biologically active in insulin-treated diabetics.

Several investigators have shown that non-diabetic subjects with atheromatous disease of coronary, cerebral and peripheral arteries have an elevated insulin response to oral glucose compared with matched controls without evidence of vascular disease [92-95]. There is only one report of a similar study among diabetics [55]. It was found that diabetics with atherosclerosis, who where not treated with insulin, had a statistically insignificant elevation of basal insulin levels and a significantly higher insulin:glucose ratio than diabetics without atherosclerosis. Diabetics with atherosclerosis were also more obese and had higher serum triglyceride levels than those without atherosclerosis and both of these abnormalities were associated with higher insulin levels, as has been previously reported in non-diabetics [60].

Some studies have suggested that insulin therapy is more common in diabetics with atherosclerosis than those without [47], and that the mortality from cardiovascular disease is higher in insulin-treated female diabetics than in those treated with diet or oral agents [35]. Other studies have reported a higher mortality in diabetics treated with oral agents than with insulin [27]. However, since there are so many variables involved in the therapy of diabetes and more severe diabetics are likely to receive more intensive therapy, no conclusions can be drawn from non-randomised retrospective studies of this type. The UGDP prospective study found no difference between the mortality from cardiovascular disease in diabetics receiving either of two insulin regimes and those receiving a placebo [48].

In view of the reported association of elevated insulin levels and atherosclerosis in non-diabetics, and of the possibility that insulin may have a pathogenic role in the development of atherosclerosis and its associated metabolic abnormalities, much more attention has to be paid to circulating insulin levels in diabetics. The role of a relative impairment of insulin secretion in the genesis of glucose intolerance must be distinguished from the possible metabolic effects of circulating insulin concentrations which, in absolute terms, are often higher than those in non-diabetic persons of ideal body weight. There is a considerable amount of evidence that many diabetics have high levels of circulating insulin of endogenous or exogenous origin. Much more investigation is required to establish the role of insulin in the pathogenesis of the vascular complication of diabetes.

Experimental models of diabetes and atherosclerosis

Attempts to produce arterial lesions in animals with experimental diabetes have been generally unsuccessful and have produced results which confuse rather than clarify the general picture. Diabetes may be produced surgically by removing most or all of the pancreas, or pharmacologically with the use of alloxan or streptozotocin which destroy the insulin-secreting cells of the pancreas, as well as having effects elsewhere in the animal. Recently, strains of spontaneously diabetic animals have been developed, most of the animals being obese and many having hyperinsulinism. The comparability of animal models to spontaneous human diabetes can be questioned.

Simultaneously in 1949, Duff and McMillan [96], and McGill and Holman [97] reported that rabbits with alloxan diabetes fed high cholesterol diets developed fewer vascular lesions than control animals fed the same diet. Later, Duff *et al.*[98] reported that in alloxanised animals, correction of hyperglycaemia with insulin reversed the inhibition of atherosclerosis. When the pancreas was temporarily excluded from the circulation during alloxan administration, the vascular disease induced by cholesterol feeding was similar to that in control animals [99]. Thus experimental ablation of insulin-secretory cells appears to protect animals from diet-induced atherosclerosis.

Much of the other experimental work on diabetes and arterial disease has produced confusing results because the animals have been given insulin [100] [101] or the circulating lipid levels in the diabetic animals have been much higher than in the controls [101] [102]. Nevertheless, in some studies, the diabetic animals developed less, or no greater lesions than controls despite higher lipid levels [103] [104]. It was later found that when serum lipid levels were equilibrated by dietary means, non-diabetic and pancreatectomised diabetic rats had similar arterial lesions, and it was suggested that the artery of the experimentally diabetic animal is not more susceptible than the non-diabetic artery at any given level of serum lipids [105]. This confirmed the earlier finding that there is a linear relationship between the plasma cholesterol and the rate of influx of cholesterol into the artery, and that the relationship is unaltered in experimental diabetes [106].

However, if the diabetic animals were treated with insulin in doses sufficient to prevent glycosuria, serum cholesterol levels were slightly reduced, but there was no reduction in the incidence or severity of vascular lesions as had occurred when similar reductions of lipids had been induced by dietary means [107]. A similar situation occurred when insulin and saline were infused respectively into the right and left femoral arteries of alloxan diabetic dogs [108]. The insulin-treated artery developed intimal and medial proliferation and contained more cholesterol and fatty acids than the control artery.

The effect of diabetes on the metabolism of the arterial wall has also been studied. Fatty acid and cholesterol synthesis were depressed in aortae from alloxan diabetic animals [109] as were glucose uptake and utilisation [110-112]. Insulin had no effect when added *in vitro,* but administration in insulin to the diabetic animals *in vivo* restored the metabolism of the aorta towards normal. The artery contains a hormone-sensitive lipase whose activity is increased by experimental diabetes and suppressed by insulin [113].

The results of the experiments described suggest that the level of circulating insulin is important in the lipid metabolism of the artery and the development of arterial lesions. It also appears that hyperglycaemia in association with absolute insulin deficiency does not enhance the development of arterial lesions. However, it is difficult to extrapolate the results of these experiments to spontaneous diabetes in man.

The atheromatous lesion

Conventionally, human atheromatous lesions are classified as fatty streaks, raised plaques and complicated lesions, according to morphological appearance. It is assumed that the lesions progress through these three stages, although of course the sequence has never actually been observed. The fact that aortic fatty streak lesions have the same prevalence in populations with a low incidence of atherosclerotic vascular disease as in populations with a high incidence of this disease [114] has cast some doubt on the importance of these lesions in the development of atherosclerosis. However, there appears to be a good correlation between the frequency of coronary artery fatty streaks and the prevalence of coronary atherosclerosis [114]. Thus the fatty streak may be capable of progression or regression, or may be independent of the more complex lesions.

Although the composition of the advanced atherosclerotic lesion is easily determined, it is only recently that the cytology of the earliest lesions have been studied. In both experimental atherosclerosis in animals [115] and spontaneous atherosclerosis in human[116-118], the earliest identifiable lesions consist of accumulation of smooth muscle cells in the arterial intima. These cells may represent proliferation of smooth muscle cells which have migrated from the media, or the proliferation of a single cell already present in the intima [119]. In more advanced lesions, the smooth muscle cells are lipid-laden and assume the form of foam cells. Ruptured foam cells surrounded by lipid and connective tissue proteins are seen in the later stages of the disease. It is assumed that the sequence starts with proliferation of smooth muscle cells, followed by accumulation of lipid and connective tissue, with eventual calcification, ulceration and superimposed thrombosis[120] (Figure 5.1.).

The lipids of the atheromatous lesion consist of cholesterol (both free and esterified), phospholipid (especially sphingomyelin) and some triglyceride [121]. Of the lipids involved, cholesterol, particularly the esterified fraction, accumulates more rapidly than the other lipids, with the result that the proportion of cholesterol in the arterial lipids increases while the proportion of other lipid decreases, despite an increase in the absolute concentration of these lipids. There is no evidence that the composition of the atheromatous lesions in diabetics differs from that in non-diabetics.

The source of the lipids of the atherosclerotic lesion remains undetermined. It is clear that the arterial wall is an active metabolic tissue

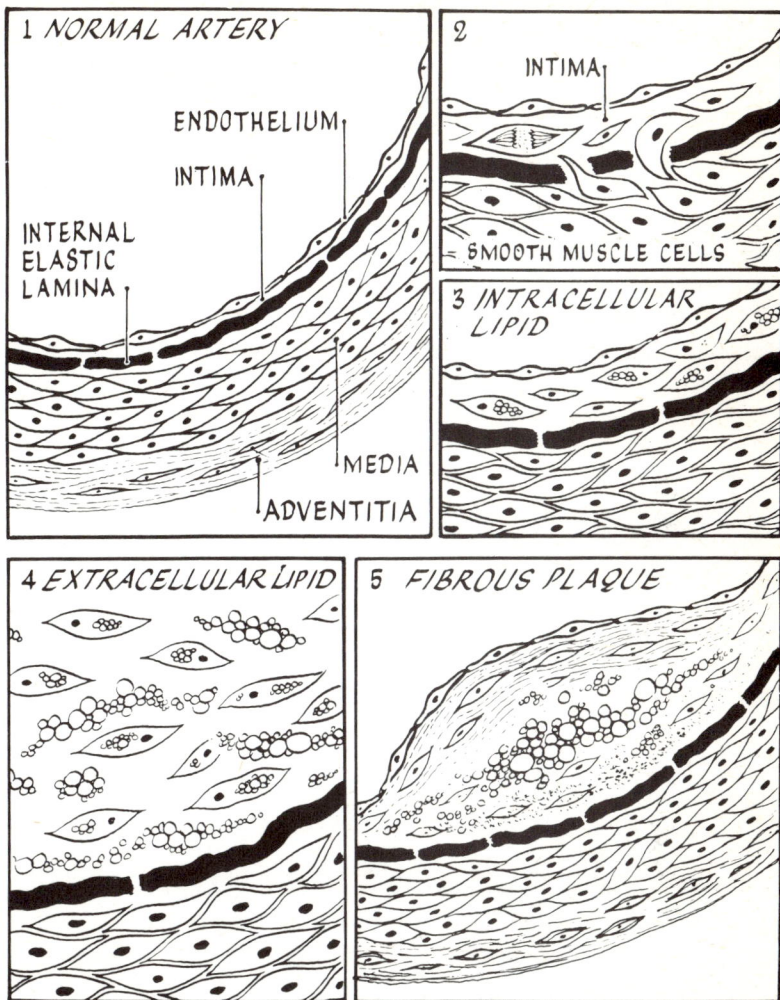

Figure 5.1 Diagrammatic representation of the development of the atherosclerotic lesion. The normal artery (1) consists of a narrow intima composed of a layer of endothelial cells supported by connective tissue which normally encompasses a small number of smooth muscle cells. In the media, which is separated from the intima by the internal elastic lamina, the only cell type is the smooth muscle cell. The adventitia consists of connective tissue and includes fibroblasts.

The earliest recognisable lesion in atherosclerosis is proliferation of smooth muscle cells in the intima (2). These cells may have been present in the intima or may migrate from the media. As the lesion develops, the smooth muscle cells become filled with lipid (3) and lipid also accumulates in the extracellular space (4). Fibrosis and calcification occur, resulting in the advanced lesion which encroaches on the lumen of the artery. (From R. and Glomset, J. A.[120] by courtesy of Dr Russell Ross).

141

and such activities as respiration and glycolysis have been extensively studied [122]. Thus the concept of atherosclerosis as the result of passive infiltration of plasma lipids into an inert conduit cannot be sustained without strong supporting evidence.

There is little doubt that cholesterol can enter the arterial intima from the plasma [123] and studies in the cholesterol-fed rabbit have suggested that almost all the cholesterol in this lesion is derived from the plasma [124]. Lipoproteins have been immunologically identified in atheromatous lesions [125] and a close relationship exists in humans between the plasma cholesterol and arterial lipoprotein cholesterol concentration [125]. Recently, autoradiographic evidence has been presented for the entry of human serum lipoproteins, labelled in their protein moiety, through the rat aortic endothelium [126], and it has been calculated that endothelial transport of lipoprotein particles could account for the accumulation of cholesterol in a normal artery with age [127].

Although the lipid content of the apparently undiseased arterial wall increases with age, the lipid composition is not the same as that in the early atheromatous lesion [125]. The major cholesterol ester fatty acid of both the ageing artery and plasma is linoleic acid, whereas that of the early lesion is oleic acid. Furthermore, the lesion contains much less lipoprotein than the adjacent artery. The amorphous lipid pools which occur under the fibrous plaques of advanced lesions consist of cholesterol esters and lipoproteins of similar composition to those of the ageing artery and the plasma [125]. It appears therefore that lipid may accumulate in the arterial wall with increasing age by an infiltration of plasma lipoproteins. However, metabolism *in situ* appears to have a role in the lipid accumulation in the fat-filled smooth muscle cells of the earliest lesions [128].

Synthesis of cholesterol has been demonstrated in the arteries in several animals [129] [130] and humans [131]. However, the cholesterol synthesis rate *in vitro* is very slow, although this may not be relevant compared to the time required for development of the lesion. Cholesterol esterification has been demonstrated in arterial tissue *in vitro* [123] and has recently been localised in the foam cells of human atheromata [132]. The artery is also capable of hydrolysing cholesterol esters but appears to be unable to catabolise the sterol [123]. Excretion of cholesterol from arteries has been little studied, but cholesterol esters seem to require hydrolysis before excretion can take place.

Thus both accumulation of plasma cholesterol and cholesterol synthesis *in situ* have been demonstrated in the arterial wall. Hydrolysis of cholesterol esters and excretion of cholesterol also occur. The relative contribution of each of these processes to the development of the atheromatous lesion remains unknown, and the extremely slow rate of accumulation of lipids in the artery makes measurement of these processes virtually impossible in the spontaneous lesion. However, the composition of the lesions and the uninvolved artery suggests that local synthesis and esterification play a part in the earliest lesions, and that plasma cholesterol is important in the accumulation of tissue cholesterol in advanced lesions and the normal ageing artery.

Fatty acid synthesis has been described in the artery [122] and is increased in the experimental atherosclerotic lesion in animals. The incorporation of fatty acids into complex lipids, particularly phospholipids, has been demonstrated [133], most of the phospholipid in the arterial wall being derived from local synthesis [124]. The metabolic pathways involved in the synthesis and catabolism of arterial phospholipids have been described [127] [134] [135] and in the human artery phospholipid synthesis has been localised to the foam cells of the lesion [132]. It has been suggested that phospholipid synthesis occurs in response to cholesterol accumulation in the artery and that phospholipid, by solubilising cholesterol, 'protects' the artery against damage by the sterol [121]. The source of arterial triglyceride, which is quantitatively relatively less important than cholesterol or phospholipid in the artery, has not been studied in detail, but incorporation of labelled precursors into this lipid has been found [133] [136] [137].

Apart from lipid, the important constituent of the atheromatous lesion is connective tissue protein. The smooth muscle cell has the capacity to synthesise collagen and elastic fibres [138] [139] and arterial smooth muscle cells in culture have been shown to incorporate labelled lysine into a protein which has the characteristics of soluble elastin, and to incorporate labelled sulphate into gylcosaminoglycans, another connective tissue matrix component [120]. The factors regulating the synthesis of these connective tissue matrix components are not understood and little is known about the mechanisms that remove these components during the healing of injured arteries.

The smooth muscle cell thus appears to have a central role in the pathogenesis of atherosclerosis. Proliferation of smooth muscle cells, perhaps stimulated by exposure to a serum constituent that has leaked through an altered endothelium, is the earliest lesion [120]. It has recently been shown that platelets, which adhere to injured endothelium, release a potent growth promoting factor [140]. Intracellular accumulation of lipid then occurs, followed by extracellular lipid and connective tissue accumulation. Both direct evidence from cultured arterial smooth muscle cells and indirect evidence from studies on whole artery preparations confirm that the smooth muscle cell is able to synthesise all the constituents of the lesion and can take up lipoproteins from its surroundings. The relative roles of local synthesis and infiltration of plasma lipids in the development of the lesion is not fully understood, but the evidence available is consistent with the idea that synthesis is important in the earliest lesion.

Recent experimental studies on the metabolism of the arterial wall

(a) The effect of lipoproteins on smooth muscle cells

The role of circulating lipid abnormalities in the genesis of atherosclerosis has received considerable attention for many years. This is prompted by the considerable epidemiological evidence linking hyperlipidaemia with atherosclerotic vascular disease and by the development of arterial lesions in animals fed high fat diets. The evidence linking diabetic vascular disease with lipid abnormalities has been reviewed.

On the cellular level two important studies link hyperlipidaemia with arterial smooth muscle cell function. It has been shown that incubating cultured primate aortic smooth muscle cells in the presence of the cholesterol-rich low density lipoprotein (LDL) results in enhanced cell proliferation [120]. LDL had a greater growth-promoting effect than high density lipoprotein, even when the concentrations of lipoprotein cholesterol were equal. The growth-promoting effect of the lipoprotein is of special interest in view of the established relation between the concentration of lipoprotein cholesterol and the extent of atherosclerosis in the population as a whole. The basis for the growth-promoting effects of lipoproteins remains to be clarified.

Other studies have shown that very low density lipoproteins and high density lipoproteins are taken up by cultured smooth muscle cells [141], the uptake being linearly related to the lipoprotein concentration in the medium. The smooth muscle cells were also able to catabolise the lipoproteins, the triglyceride presumably being hydrolysed within the smooth muscle cell while the cholesterol remained, perhaps to be linked to elastin [142]. Furthermore, very low density lipoprotein 'remnants', particles which are formed in the circulation by lipolysis of most of the lipoprotein triglyceride of very low density lipoproteins and chylomicrons, are avidly taken up by cultured smooth muscle cells [143]. Electronmicroscopy revealed that the radioactive label was localised to the cellular cytoplasm and, in some cells, seen over lysosomes. The cells appeared to have a limited capacity to catabolise the remnants despite avid uptake. Thus, if remnants are normally present, if transiently, during the metabolism of very low density lipoproteins and present in increased concentrations in certain hyperlipoproteinaemic subjects, the results of these studies may provide a mechanism whereby hypertriglyceridaemia can influence the pathogenesis of atherosclerosis [144].

(b)   The polyol pathway

Polyols, or sugar alcohols, are organic compounds containing multiple alcoholic groups that are derived from sugars by the reduction of their free aldehyde or ketone groups. The polyol pathway catalyses the reduction of D-glucose to its polyol derivative, sorbitol and its subsequent oxidation to D-fructose by the following reactions:

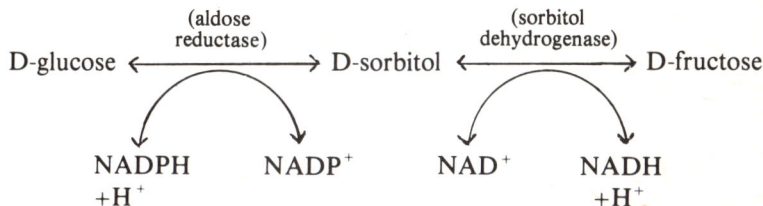

$$\text{D-glucose} \underset{\substack{\text{NADPH}\\ +\text{H}^+}}{\overset{\substack{\text{(aldose}\\ \text{reductase)}}}{\longleftrightarrow}} \text{D-sorbitol} \underset{\substack{\text{NAD}^+}}{\overset{\substack{\text{(sorbitol}\\ \text{dehydrogenase)}}}{\longleftrightarrow}} \text{D-fructose}$$

D-glucose ⟷ (aldose reductase) D-sorbitol ⟷ (sorbitol dehydrogenase) D-fructose

NADPH +H$^+$ NADP$^+$ NAD$^+$ NADH +H$^+$

These reactions appear to operate irreversibly in intact tissues and are not dependent on the presence of insulin. Aldose reductase has a low affinity for glucose and the rate of sorbitol formation therefore increases with increasing glucose concentration over a wide range. In diabetes the availablity of a large pool of free intracellular glucose in certain tissues causes increased formation of sorbitol. Polyols penetrate cell membranes poorly. Once formed, they are trapped intracellularly, the only disposition being either conversion to the respective ketosugar or slow leakage from the cell. Accumulation of polyols within the cell results in intracellular hypertonicity. Activity of the polyol pathway has been demonstrated in the lens of the eye and in peripheral nerves and has been implicated in the pathogenesis of diabetic cataract and diabetic neuropathy [145].

Recently the polyol pathway has been demonstrated in the aortic intima and media of humans and rabbits [146] [147]. Extracts of whole rabbit aorta, consisting mainly of smooth muscle cells, were found to contain sorbitol in a concentration five to six times that in normal rabbit plasma. With higher medium glucose concentrations, an increased concentration of sorbitol and fructose was found in the aortic wall and increased release of fructose, but not sorbitol into the medium occurred, suggesting increased turnover of sorbitol in the aorta. This was accompanied by an increased water content in the artery wall, most of which appeared to be intracellular. Oxygen uptake of aortic samples incubated in the higher concentrations of glucose were less than in samples incubated in lower glucose concentrations and the defect in oxygen uptake in the higher glucose concentrations could be reversed by adding mannitol to the incubation medium. Although the data were collected in *in vitro* experiments, freshly prepared aortic segments from alloxan diabetic animals showed a greater water content and a reduced rate of respiration compared with tissue from normal controls prepared under the same conditions.

It is suggested that high ambient glucose concentrations cause an increase in the activity of the polyol pathway in the arterial wall, with intracellular accumulation of sorbitol in the wall. The osmotic activity of the sorbitol and other as yet unidentified factors, causes an increase in the tissue water content, which impairs oxygen diffusion, resulting in a decrease in the respiratory rate. This may alter the metabolism of the arterial wall and possibly contribute towards the production of arterial lesions. Although the polyol pathway provides a link between hyperglycaemia and metabolism of the arterial wall, its role in the pathogenesis of atherosclerosis in diabetes is not yet clear.

(c) *Insulin activity on the artery*

The relationship of insulin secretion to atherosclerotic disease in non-diabetic subjects, and abnormalities of insulin secretion in diabetes and obesity have already been discussed. It seems that a proportion of diabetics have high levels of circulating insulin at some stage of their disease, whether this is endogenous insulin related to obesity, or due to therapeutic adminstration of insulin.

The possibility that elevated concentrations of insulin, or normal insulin concentrations at inappropriate times, may be directly related to the arterial disease has been explored. It has been found that cultured arterial smooth muscle cells proliferate when exposed to insulin [148]. This response was found with insulin concentrations within the physiological range and the proliferative response was linearly related to the insulin concentration. The response was lost in cells that had aged *in vitro* and appeared to be mediated by the cyclic 3',5'-adenosine monophosphate system.

Studies on the effect of insulin on lipid metabolism in cultured arterial smooth muscle cells have not been performed to date. However, studies on whole rat aortae, composed almost entirely of smooth muscle, have shown that when insulin was injected into the animals *in vivo*, glucose and acetate incorporation into aortic lipids was enhanced [149] [150]. Furthermore, when endogenous insulin secretion was reduced by administration of streptozotocin, the incorporation of glucose into aortic lipids *in vitro* was depressed and was directly related to the circulating insulin concentration in the animal [151].

In non-diabetic animals, administration of insulin [152] prevented the regression of arterial lesions in chickens which normally occurs when a high cholesterol diet is replaced by a normal diet. Insulin also prevented estrogen-induced protection of coronary atherosclerosis in cholesterol-fed cockerels [152]. There are two reports of the development of arterial lesions in hyperinsulinaemic laboratory animals fed a normal diet. When chickens were injected with long-acting insulin for nineteen weeks they developed significantly more lipid-containing lesions of their aortae than control birds injected with a bland solution [153]. The insulin-treated chickens also developed elevated serum triglyceride and cholesterol levels without excessive weight gain[154]. The spiny mouse *(Acomys cahirinus)* which develops obesity, diabetes and hyperinsulinism, has been reported to develop spontaneous lesions of its coronary arteries on a normal diet [155].

There is thus evidence that insulin has a direct effect on the artery which may be relevant to atherosclerosis. As is discussed elsewhere in this chapter, insulin also has an important role in the regulation of plasma lipid transport. The exact role of insulin-mediated changes in lipid metabolism in the development of atherosclerosis remains to be elucidated [156].

Conclusions

Atherosclerosis is more common in diabetics than in the general population. Although hypertension, hyperlipidaemia and obesity are all common in diabetics and probably contribute to the acceleration of atherosclerosis in diabetes, they cannot completely explain the excess experience of arterial disease in diabetes. Atherosclerosis is not closely related to the severity or duration of diabetes, nor is it prevented by reduction of blood sugar by therapeutic means. It is probable that several factors contribute towards the development of atherosclerosis in diabetes, but whether these factors are all environmental or include some abnormality inherent to the diabetic state is not clear. The fact that diabetics can survive with their disease for long periods without developing vascular disease [157], and that a low incidence of atherosclerosis is found in diabetics in some parts of the world [39], suggests that environmental factors are important. However, the particular factor involved remains to be identified.

The important cell in the atherosclerotic lesion is the smooth muscle cell and proliferation of smooth muscle cells is the earliest identifiable lesion in atherosclerosis. The smooth muscle cell may therefore be the final common pathway through which many factors influence the development of the lesion. The arterial smooth muscle cell may be considered to respond to a large number of stimuli with a limited number of responses [158]—proliferation, synthesis of connective tissue, synthesis and inbibition of lipid (Figure 5.2). Thus the various physiological derangements which occur in diabetes may act on the cell to produce a lesion which in the fully developed form is indistinguishable from the lesion in non-diabetics.

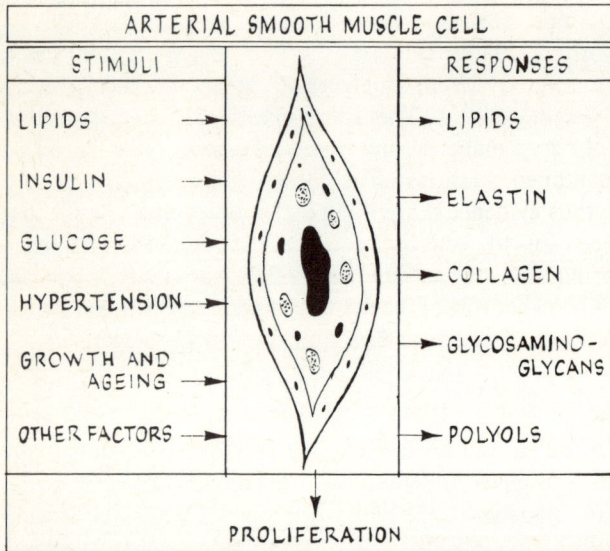

Figure 5.2 Diagrammatic representation of the possible role of the arterial smooth muscle cell in the development of atherosclerosis in diabetes. A variety of stimuli act on the smooth muscle cell: the cell has a limited number of responses–proliferation and synthesis or inhibition. Thus the arterial smooth muscle cell may act as a common pathway by which many different factors influence the development of atherosclerosis. (From Wolinsky, H.[158] by courtesy of Dr. Harvey Wolinsky).

LIPID METABOLISM IN DIABETES

Diabetes mellitus is often associated with an excessive accumulation of one or more of the major lipids transported in plasma. For clinical purposes, this hyperlipidaemia may manifest as hypertriglyceridaemia, hypercholesterolaemia or both. Thus, levels of the lipid-carrying molecular aggregates – the lipoproteins – are elevated (hyperlipoproteinaemia). Although the lipoproteins are a continuum of size and composition, they have been divided into the triglyceride-rich, very low density (pre-beta) lipoproteins and the cholesterol-rich, low density (beta) lipoproteins. In addition chylomicrons, the largest and most triglyceride-rich of the lipoproteins, enter the circulation from the gut and may accumulate in plasma in certain diabetics, causing turbidity or lactescence.

The association of lactescent serum with diabetics was first noted in 1799 by Mariet, and in the pre-insulin era lactescent serum was observed commonly in diabetic subjects [159]. Man and Peters found that triglyceride was the primary lipid to be elevated in diabetes [160]. More recently, elevations in triglyceride-rich very low density lipoproteins have been found to be a common abnormality [161]. However, chylomicronaemia, while not as prevalent, may be observed in diabetics with recent onset of ketoacidosis or with chronic untreated fasting hyperglycaemia [162] [163]. Increased plasma cholesterol levels also occur in some diabetic patients [46] [164]. In some patients this increase in cholesterol occurs with elevation of low density lipoproteins with normal levels of very low density lipoproteins [164]. However, hypercholesterolaemia also commonly occurs in association with marked elevations in levels of the triglyceride-rich very low density lipoproteins since cholesterol is also contained in these lipoproteins.

Overt diabetes with an elevation of fasting glucose levels and the absence of an acute insulin response to intravenous glucose [165] will be considered separately from the postprandial glucose intolerance associated with normal fasting glucose levels often co-existing with obesity, uraemia and other forms of peripheral insulin resistance. This form of glucose intolerance, characterised by hyperinsulinism and relative insulin deficiency with peripheral insulin resistance [84], may or may not proceed to the complete diabetic syndrome.

Hypertriglyceridaemia

The reported prevalence of hypertriglyceridaemia in diabetic populations is variable. Up to 30% of patients followed in diabetes clinics have elevated plasma triglyceride levels [76] [166]. Untreated fasting hyperglycaemia is frequently associated with hypertriglyceridaemia and plasma triglyceride levels are reversed to or towards normal with therapy aimed at reduction of the elevated glucose levels [159] [162]. Glucose intolerance with normal fasting glucose levels is also frequently a concomitant of hypertriglyceridaemia. It has been suggested that obesity plays a part both in the increased triglyceride levels and the glucose intolerance seen in obese, adult-onset, mildly diabetic patients [167] since weight reduction leads to reversal of both glucose intolerance and hypertriglyceridaemia.

The mechanism of hypertriglyceridaemia in untreated overt diabetes could either be an increase in splanchnic production of triglyceride-rich very low density lipoproteins, or a defect in the removal of the very low density lipoproteins and/or chylomicrons.

(a) Triglyceride production

Increases in very low density lipoprotein production might be expected in the untreated diabetic state since there are marked increases in the two major precursors for triglyceride synthesis — plasma free fatty acids and glucose. With insulin deficiency there is a marked increase in free fatty acid mobilisation from adipose tissue triglyceride stores because of the uninhibited activity of adipose tissue hormone-sensitive lipase. Any decrease in plasma volume in the untreated diabetic state will lead to an increase in circulating plasma catecholamine levels [168] and further enhance free fatty acid mobilisation. Abnormally high and non-suppressed glucagon and growth hormone levels in diabetes potentially add still further to uninhibited free fatty acid mobilisation.

When high concentrations of free fatty acids are presented to the liver in insulin deficiency, some of the fatty acids enter hepatic cell mitochondria, undergo oxidation to acetyl-coenzyme A and then progress to ketone formation or further oxidation to carbon dioxide. A certain proportion of the free fatty acids remains in the cytoplasm of the hepatic cell and, with α-glycerol phosphate, is esterified to triglyceride. In the diabetic state associated with ketosis, the proportion of the fatty acids removed from the plasma by the liver remains the same as in the normal fasted or fed state. Thus, in the face of increased free fatty acid turnover, there is increased absolute uptake of free fatty acids by the liver. For several hours after the development of insulin deficiency, there is an increase in free fatty acid esterification to triglyceride in absolute terms [169], although the relative proportion of free fatty acids being esterified might be decreased [170]. This may explain in part the increase in hepatic triglyceride (fatty liver) noted in patients who die in ketoacidosis[171]. However, with longer periods of insulin deficiency, there is a decrease in fatty acid esterification and triglyceride secretion by the liver [172] [173] despite continuing hypertriglyceridaemia, suggesting that with prolonged insulin deficiency the cause of the increase in plasma triglyceride levels is a defect in plasma triglyceride removal. Thus, in the untreated diabetic, blockade of fatty acid mobilisation with nicotinic acid lowers plasma triglyceride levels during acute insulin deficiency [174], but has no effect on plasma triglyceride levels with longer periods of insulin deficiency[175] even though plasma free fatty acid and ketone levels are decreased.

An increase in triglyceride production from glucose seems unlikely since hepatic lipogenesis is minimal in the fasted state and contributes little to triglyceride synthesis in the fed state when associated with insulin deficiency [176].

(b)   Triglyceride removal

On the other hand, defects in plasma triglyceride removal mechanisms have been demonstrated by several techniques in diabetes and support the indirect evidence for a defect in the catabolism of plasma triglyceride. Measurement of plasma triglyceride transport rates by estimation of the clearance of triglyceride-rich lipoproteins endogenously labelled with radioglycerol have been performed in untreated and treated diabetic patients [177]. Most of the diabetic subjects had high transport rates indicative of overproduction when compared to a normal population, while some patients had overt removal defects. However, when the diabetic patients with uncontrolled hyperglycaemia (fasting glucose above 140 mg/dl)[177] are compared to patients with glucose levels below 140 mg/dl[177] and non-diabetic normal [178] and hypertriglyceridaemic patients [179], one-third of the severely hyperglycaemic patients appeared to have a plasma triglyceride removal defect. The role of primary and other secondary causes of hypertriglyceridaemia in those studies was not evaluated. A defect in the removal rate of exogenously administered fat (Intralipid) in untreated diabetics was not demonstrated when compared to that in a group of non-diabetic hypertriglyceridaemia patients. However, when the removal rate of Intralipid was measured in the same patient before and after diabetic therapy [180] an increase in the removal rate was noted which was associated with a decrease in the plasma triglyceride level. However, as the quantity of circulating triglyceride may affect the removal rate [181], a more rapid removal rate need not imply an improvement in a primary removal defect, but may simply reflect lower plasma triglyceride levels. On the other hand, an increase in the maximal removal capacity of Intralipid was demonstrated with diabetic therapy and this can be interpreted as evidence for correction of a plasma triglyceride removal defect.

The specific defect in plasma triglyceride removal appears to relate to the enzyme lipoprotein lipase located at or near the capillary endothelium. This enzyme normally removes triglyceride from circulating triglyceride-rich lipoproteins by hydrolysis of the triglyceride into free fatty acids and glycerol [182]. The free fatty acids are then taken up by adipose tissue and the liver and re-esterified to triglyceride or utilised directly for energy by oxidation, mainly in muscle.

Rarely, in severe, overt, chronically untreated diabetes, a defect in lipoprotein lipase occurs that may be similar to the defect found in familial lipoprotein lipase deficiency. This can be demonstrated as a 'deficiency' of post-heparin lipolytic activity (PHLA) in plasma following a low (10 units/kg) dose of intravenous heparin. This abnormality is corrected rapidly towards normal with insulin therapy with a concomitant decrease in

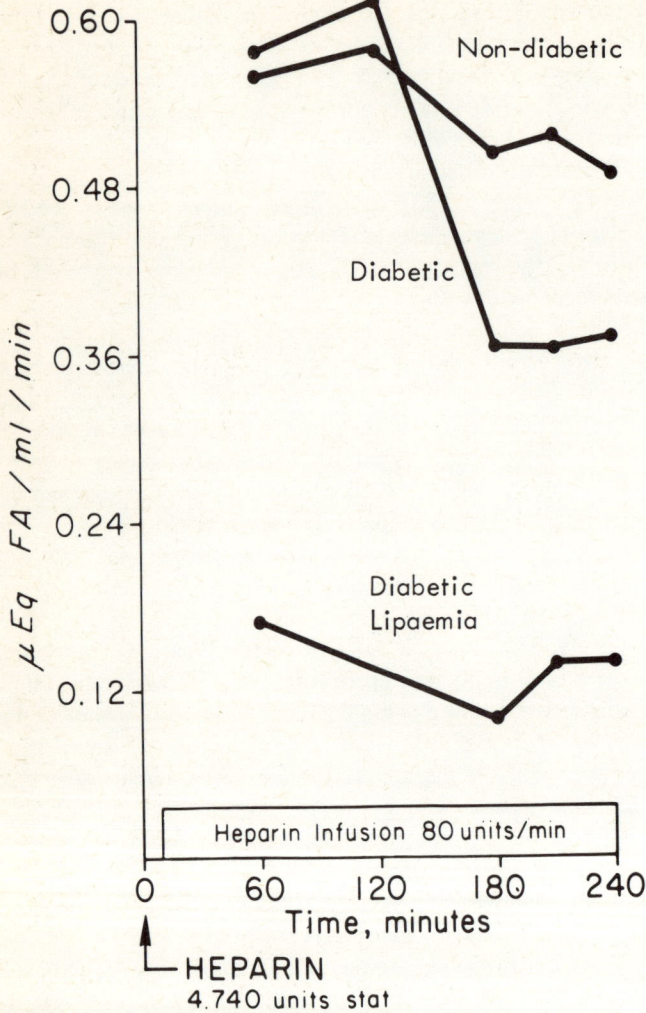

Figure 5.3 Post-heparin lipolytic activity during a prolonged heparin infusion in non-diabetic and diabetic subjects. In non-diabetics, PHLA levels remain elevated throughout the four hours of the infusion. In subjects with severe diabetes and marked lipaemia (diabetic lipaemia), PHLA levels depressed throughout the infusion. In subjects with moderately severe diabetes and hypertriglyceridaemia, PHLA is elevated normally during the early part of the infusion but falls during the latter part of the infusion.

plasma triglyceride levels [162]. However, most untreated diabetics do not have such a demonstrable abnormality [183], but still have a decrease in plasma triglyceride following therapy for hyperglycaemia. Recently, in such individuals a more subtle defect in post-heparin lipolytic activity has been demonstrated during a high dose (60 units/kg/h) prolonged heparin infusion [184]. In these subjects, post-heparin lipolytic activity during the late phase of the heparin infusion (from 3–6h) is diminished, even though the PHLA is normal during the early phase of this heparin infusion (Figure 5.3). Higher fasting glucose levels were associated with greater late drops in PHLA. The relationship between the degree of hyperglycaemia and the late phase PHLA suggests this defect may be related to insulin deficiency. Following several months of insulin therapy the defect in PHLA returns to normal with a concomitant decrease in plasma triglyceride levels. For reasons that are not yet understood, the abnormality does not immediately correct following insulin treatment, but instead persists until after two to three months of therapy. A similar biphasic release of lipoprotein lipase from the hearts of rats perfused with heparin has been found[185]. In diabetic rat hearts there was a defect in the release of lipoprotein lipase during the second phase of the infusion, which was returned to normal by insulin therapy.

It now appears that part of the defect in PHLA during the heparin infusion is related to an abnormality in adipose tissue lipoprotein lipase in diabetic animals[186][187] and in man [188]. In untreated diabetes the levels of adipose tissue lipoprotein lipase are lower than in non-diabetics with normal or elevated triglyceride levels [188]. Adipose tissue lipoprotein lipase appears to be normal in treated diabetics [189]. Further support for the role of insulin deficiency in the development of the abnormality in lipoprotein lipase and hypertriglyceridaemia is derived from studies in which insulin therapy was withdrawn from juvenile, ketosis-prone diabetic patients [190]. After forty-eight hours of insulin deprivation, plasma triglyceride levels increased in association with a decrease in postheparin lipolytic activity.

Familial hypertriglyceridaemias and diabetes
Part of the problem in the evaluation of triglyceride transport in diabetes mellitus is the observation that in some diabetics, elevated plasma triglyceride levels will persist even after therapy for hyperglycaemia which will correct the abnormality in lipoprotein lipase. In this persistent hypertriglyceridaemia due to diabetes itself or to some other cause of hypertriglyceridaemia? In some of these individuals, not enough time may have elapsed since the last episode of ketoacidosis or from the time of initiation of antihyperglycaemic therapy for correction of the abnormality

in lipoprotein lipase and the resultant hypertriglyceridaemia. In others, familial forms of hypertriglyceridaemia may co-exist, since their presence has been suggested to be frequent in populations of diabetic patients. In addition, other secondary causes of hypertriglyceridaemia frequently occur in patients with diabetes mellitus. One or more of these factors may account for the persistent elevation of plasma triglyceride levels.

Diabetes mellitus and familial forms of hypertriglyceridaemia appear to occur together more frequently than would be expected if they were independent disorders [191]. However, it is important to know if this association is a real one, or if it is due to the preferential selection of patients by the physician either because they are more symptomatic, or because the presence of either an elevation of plasma glucose or triglyceride level leads to the measurement of the other circulating metabolite. Recent studies to examine this question suggest that the association of overt diabetes and monogenic forms of hypertriglyceridaemia [192] (familial combined hyperlipidaemia and 'pure' familial hypertriglyceridaemia) is more apparent than real [193]. When the adult relatives of patients with both overt diabetes and a familial form of hypertriglyceridaemia were examined, the prevalence of diabetes was increased in the relatives, but was the same in the normolipidaemic relatives as in the hyperlipidaemic relatives. Since the hyperlipidaemic relatives did not have more diabetes, these two disorders must segregate and be inherited independently. The index patients of these families often present to the physician because of symptoms related to hypertriglyceridaemia, such as abdominal pain. In these subjects the defect in lipoprotein lipase related to untreated diabetes interacted with the familial lipid disorder and the combination resulted in marked hypertriglyceridaemia. This can be demonstrated in two patients with overt PHLA deficiency which corrected with insulin therapy. In spite of similar defects in lipoprotein lipase, one patient had a plasma triglyceride level of 4553 mg/dl and the other only 266 mg/dl. Following insulin therapy, the first patient had a fall in plasma triglyceride to 453 mg/dl. Examination of her relatives revealed concomitant familial hypertriglyceridaemia. The second patient had a fall in plasma triglyceride levels to normal (119 mg/dl) and plasma lipids were normal in five adult first degree relatives examined, indicating that she probably did not have a concomitant familial lipid abnormality. Similar interactions with familial lipid disorders were found in patients with the more subtle defect in lipoprotein lipase (late phase PHLA deficiency – see page 151). Their plasma triglyceride levels fell after several months of antidiabetic therapy to levels found in their non-diabetic hypertriglyceridaemic relatives [184].

Secondary hypertriglyceridaemias and diabetes

Several secondary causes of hypertriglyceridaemia, other than insulin deficiency, are known to occur in patients with hyperglycaemia. Two such disorders, obesity and uraemia, lead to postprandial glucose intolerance [194] or may progress to fasting hyperglycaemia. However, they also are associated with hypertriglyceridaemia without glucose intolerance [167 195]. Whether or not these patients have 'true' diabetes is difficult to determine since, at this time, there is no specific genetic marker for diabetes.

In the majority of diabetic patients the disease is first diagnosed in adult life and most of these patients are obese. Obesity itself is known to be associated with elevated plasma triglyceride levels, presumably through overproduction of triglyceride [196] by the liver and gastrointestinal tract. Plasma triglyceride production appears to be modulated directly be plasma insulin levels [197 198] which are increased in obesity and possibly inhibited by plasma glucagon [199]. The increase in basal (and stimulated) insulin levels associated with obesity appears to be due to a pancreatic response to compensate for decreased insulin sensitivity—insulin resistance—associated with obesity. Transient weight gain and weight loss are associated with parallel increases and decreases in triglyceride and insulin levels. Furthermore, patients restudied at new lower body weights after weight loss have lower insulin and triglyceride levels with lower very low density triglyceride production rates[200].Since glucose levels respond similarly to weight changes, it has been suggested that the frequent association between increased glucose and triglyceride levels in such subjects is due to the common aggravating factor—obesity[167 201].Chronic renal failure with uraemia is also known to lead to hypertriglyceridaemia[195] and occurs in diabetes. Insulin resistance in uraemia and perhaps decreased insulin degradation in the diseased kidney result in elevated insulin levels leading to enhanced splanchnic triglyceride production. It has also been suggested that uraemia may be associated with a triglyceride removal defect manifested as an abnormality in lipoprotein lipase.

Another abnormality associated with hypertriglyceridaemia [202] that occurs in diabetes is the nephrotic syndrome. Again, the hypertriglyceridaemia in this disorder might be due to overproduction of plasma triglyceride, or there may be a lipoprotein lipase-related triglyceride removal defect. The nephrotic syndrome is also commonly associated with hypercholesterolaemia and may in part cause the increased cholesterol levels seen in certain diabetic patients.

The response of fasting plasma triglyceride to high carbohydrate diets ('carbohydrate induction') in treated diabetic patients or in milder diabetics does not appear to be any different than in normal subjects or in non-diabetic hypertriglyceridaemic patients[78] [203].

Hypercholesterolaemia

Far less is known about cholesterol transport in diabetes. Familial hypercholesterolaemia either as the monogenic or the polygenic disorder [192], does not appear to be more frequent in diabetes than in the general population [204]. Hypercholesterolaemia, possibly due to secondary abnormalities, has been reported in certain groups of diabetic patients. In some studies, young patients with diabetes have been found to have an elevated plasma cholesterol without concomitant hypertriglyceridaemia [40]. In a recent report, elevated plasma cholesterol was found more often in insulin-treated patients than in oral sulphonylurea-treated patients or non-diabetic controls [205]. These insulin-treated patients also tended to be a younger group of individuals. The reason for the increased plasma cholesterol in these patients is unknown at the present time. Perhaps hypercholesterolaemia results in part from the use of carbohydrate-restricted, high fat diets emphasised particularly in insulin-treated diabetic populations. This possibility has to be examined. The insulin-treated group of subjects are the ones most likely to live long enough with diabetes to develop renal disease with development of the nephrotic syndrome. The role of the hypercholesterolaemia of the nephrotic syndrome in the increased prevalence of hypercholesterolaemia in insulin-treated diabetics has yet to be determined.

Complications of hyperlipidaemia

The premature atherosclerosis that occurs in patients with diabetes and glucose intolerance has been discussed. Increased plasma triglyceride and cholesterol concentrations contribute to the multiplicity of factors that predispose the diabetic patient to accelerated atherogenesis. The enhanced risk with hyperlipidaemia appears to be continuous with increasing elevations of triglyceride and cholesterol levels.

Another not uncommon complication of marked hypertriglyceridaemia is abdominal pain or pancreatitis [206]. In diabetic patients this occurs almost exclusively during episodes of uncontrolled hyperglycaemia with or without ketosis. Although all abdominal pain that occurs in these situations may not be related to concomitant hypertriglyceridaemia, these two disorders occur together with unexpected frequency. It appears that the abdominal pain is associated with the presence of circulating fat particles of dietary origin ('chylomicrons'), and that manoeuvres which lower triglyceride levels and lessen chylomicronaemia lead to amelioration of the abdominal pain. It seems to follow the same course as eruptive xanthomata ('xanthomata diabeticorum') which reflect chronic chylomicronaemia [207] and are a type of skin lesion which was much more frequent in diabetic patients before the use of insulin [159].

In order for triglyceride levels to be high enough for chylomicrons to persist in plasma after an overnight fast there often needs to be a combination of more than one cause of hypertriglyceridaemia. Exceptions to this generalisation include familial lipoprotein lipase deficiency, broad beta disease and perhaps lipoprotein lipase abnormalities associated with abnormalities in gamma globulins. To have plasma triglyceride levels high enough to cause pancreatitis, an interaction between several forms of hypertriglyceridaemia is almost always found [206].

In diabetic patients, marked hypertriglyceridaemia and abdominal pain during uncontrolled diabetes is often, if not always, related to an interaction of an abnormality in lipoprotein lipase with an underlying familial form of hypertriglyceridaemia. Possibly, the marked hypertriglyceridaemia might be due to a combination of untreated diabetes and some other secondary form of hypertriglyceridaemia, such as oestrogen therapy or excessive alcohol intake. In any case, once pancreatitis occurs, whatever insulin reserve remains becomes compromised and a more severe degree of insulin deficiency occurs, further aggravating the abnormality in lipoprotein lipase. Since the abnormality in lipoprotein lipase takes time to be corrected with insulin therapy, these patients are often difficult to treat. Recurrent pancreatitis may occur when dietary fat intake is increased. Several months of low-fat intake associated with long-term clofibrate therapy will lower triglyceride levels and can prevent recurrence of this abdominal pain [206].

Congenital total lipodystrophy
Congenital total lipodystrophy, a form of lipoatrophic diabetes, offers an interesting model for the study of the interrelationships among insulin resistance, insulin deficiency and plasma triglyceride metabolism. This disorder is an autosomal recessive disease characterised by the absence of adipose tissue mass noted from birth. Insulin resistance is noted early, with early development of glucose intolerance, followed by fasting hyperglycaemia which is quite resistant to insulin therapy. Ketoacidosis is rare and this disorder is somewhat similar to the type of diabetes which begins in adults and is associated with obesity and a less severe form of insulin resistance.

Congenital total lipodystrophy is associated with hypertriglyceridaemia, which is often marked. Before the onset of fasting hyperglycaemia, some of the increase in plasma triglyceride levels may be related to the peripheral insulin resistance by a mechanism similar to that which occurs with obesity.

Post-heparin lipolytic activity at this time is normal. Many very young patients, however, have marked hypertriglyceridaemia, often transient, which occurs prior to the development of symptomatic diabetes. The reason for this is highly speculative, but a possibility is suggested by a patient who developed marked lactescence of his serum at six months of age during a period of growth often characterised by an increase in the deposition of adipose tissue stores[208]. This patient may have developed transient hypertriglyceridaemia because of difficulty in deposition of fat in his adipose tissue.

After the development of fasting hyperglycaemia, overt PHLA deficiency is common and can easily account for the hypertriglyceridaemia common to this phase of lipodystrophy [209]. This disorder is different from the PHLA deficiency seen in severe, prolonged untreated hyperglycaemia, in that PHLA deficiency does not readily respond to insulin therapy. This may be due to the severe insulin resistance that occurs in this disorder and thus, adequate insulinisation may never occur. An alternative explanation may be that the adipose tissue cell that lacks stored triglyceride loses its innate ability to synthesise lipoprotein lipase.

Conclusions

Hypertriglyceridaemia is common in untreated or decompensated diabetes. Very early after the onset of insulin deficiency there may be an increase in splanchnic triglyceride production because of a marked influx of plasma free fatty acids, resulting in an increase in hepatic lipid content and circulating triglyceride-rich, very low density lipoproteins. With longer periods of insulin deficiency, triglyceride secretion decreases and a defect in adipose tissue lipoprotein lipase develops, causing impaired plasma triglyceride removal and continuing hypertriglyceridaemia.

Marked hypertriglyceridaemia with pancreatitis and eruptive xanthomata associated with chylomicronaemia may occur when insulin deficiency is superimposed on an underlying familial lipid disorder.

Following insulin therapy the defect in lipoprotein lipase is corrected to normal. Persistent hypertriglyceridaemia may be due to an independent familial lipid disorder or to the common secondary causes of hypertriglyceridaemia present in diabetic populations, such as obesity, uraemia, or the nephrotic syndrome.

Although less common, hypercholesterolaemia is also found in some groups of diabetic patients. The reason for this hypercholesterolaemia is speculative at present.

References

1. Marks, H. H. (1965). Longevity and mortality of diabetes. *Amer. J. Public Health,* **55,** 416
2. Hayward, R. E. and Lucena, B. C. (1965). An investigation into the mortality of diabetics. *J. Inst. Actuaries,* **91,** 286
3. National Heart and Lung Institute (1971). *Arteriosclerosis—a report by the National Heart and Lung Institute Task Force on Arteriosclerosis* (Bethesda, Maryland: National Institutes of Health)
4. Siperstein, M. D. (1973). Diabetic microangiopathy: electron microscope findings in human muscle. *Adv. Metab. Dis.* **(Suppl. 2),** 349
5. Strandness, D. E., Priest, R. E. and Gibbons, G. E. (1964). Combined clinical and pathologic study of diabetic and nondiabetic peripheral arterial disease. *Diabetes,* **13,** 366
6. Oakley, W. G., Pyke, D. A. and Taylor, K. W. (1973). *Diabetes and its Management* (Oxford: Blackwell)
7. American Diabetes Association (1969). Standardization of the oral glucose tolerance test. *Diabetes,* **18,** 299
8. Blottner, H. (1930). Coronary disease in diabetes mellitus. *New Engl. J. Med.,* **203,** 709
9. Nathanson, M. H. (1932). Coronary disease in 100 autopsied diabetics. *Amer. J. Med. Sci.,* **183,** 495
10. Robbins, S. L. and Tucker, A. W. (1944). The cause of death in diabetes. *New Engl. J. Med.,* **231,** 865
11. Stearns, S., Schlesinger, M. J. and Rudy, A. (1947). Incidence and clinical significance of coronary artery disease in diabetes mellitus. *Arch. Int. Med.,* **80,** 463
12. Clawson, B. J. and Bell, E. T. (1949). Incidence of fatal coronary disease in non-diabetic and diabetic persons. *Arch. Pathol.,* **48** 105
13. Liebow, I. M., Hellerstein, H. K. and Miller, M. (1955). Arteriosclerotic heart disease in diabetes mellitus. *Amer. J. Med.,* **18,** 438
14. Breithaupt, D. J. and Leckie, R. B. (1961). Diabetes mellitus: a study of the practicability of life insurance. *Can. Med. Ass. J.,* **85,** 299
15. Levine, S. A. (1929). Coronary thrombosis: its various clinical features. *Medicine,* **8,** 245
16. Master, A. M., Dack, S. and Jaffe, H. L. (1939). Age, sex and hypertension in myocardial infarction due to coronary occlusion. *Arch. Int. Med.,* **64,** 767
17. Aarseth, S. (1953). Cardiovascular renal disease in diabetes mellitus. *Acta Med. Scand.,* **(Suppl.),** 281
18. Bartels, C. C. and Rullo, F. R. (1958). Unsuspected diabetes mellitus in peripheral vascular disease. *New Engl. J. Med.,* **259,** 633
19. Sievers, J., Blomqvist, G. and Biorck, G. (1961). Studies on myocardial infarction in Malmo, 1935 to 1954, VI. Some clinical data with particular reference to diabetes, menopause and heart rupture. *Acta Med. Scand.,* **169,** 95
20. Fabrykant, M. and Gelfand, M. L. (1964). Symptom-free diabetes in angina pectoris. *Amer. J. Med. Sci.,* **247,** 665
21. Conrad, M. C. (1967). Large and small artery occlusion in diabetics and non-diabetics with severe vascular disease. *Circulation,* **36,** 83
22. Datey, K. K. and Nanda, N. C. (1967). Hyperglycemia after acute myocardial infarction. Its relation to diabetes mellitus. *New Engl. J. Med.,* **276,** 262

23. Bailey, R. R. and Beavan, D. W. (1968). Diabetes mellitus and myocardial infarction. *Austr. Ann. Med.,* **17,** 312

24. Wood, P. (1968). *Diseases of the Heart and Circulation* (London: Eyre and Spottiswoode)

25. Pyke, D. A. (1968). Coronary disease and diabetes. *Postgrad. Med. J.,* **44,** 966

26. Jakobson, T., Kahanpaa, A. and Maenpaa, V. J. (1968). Prednisone-glucose tolerance and serum lipids in survivors of myocardial infarction. *Acta Med. Scand.,* **184,** 451

27. Soler, N. G., Pentecost, B. L., Bennett, M. A., Fitzgerald, M. G., Lamb, P. and Malins, J. M. (1974). Coronary care for myocardial infarction in diabetics. *Lancet,* **i,** 475

28. Mitchell, J. R. A. and Schwartz, C. J. (1965). *Arterial Disease* (Oxford: Blackwell)

29. Wahlberg, F. and Thomasson, B. (1968). Glucose tolerance in ischaemic cardiovascular disease. In: *Carbohydrate Metabolism and its Disorders* (F. Dickens, P. J. Randle and W. J. Whelan, editors) (London: Academic Press)

30. Wahlberg, F. (1966). Intravenous glucose tolerance in myocardial infarction, angina pectoris and intermittent claudication. *Acta Med. Scand.,* **180, (Suppl. 453)**

31. Heinle, R. A., Levy, R. I., Fredrickson, D. S. and Gorlin, R. (1969). Lipid and carbohydrate abnormalities in patients with angiographically documented coronary artery disease. *Amer. J. Cardiol.,* **24,** 178

32. Falsetti, H. L., Schnatz, J. D., Greene, D. G. and Bunnell, I. L. (1970). Serum lipids and glucose tolerance in angiographically proved coronary artery disease. *Chest,* **58,** 111

33. Mainland, D. (1953). The risk of fallacious conclusions from autopsy data on the incidence of diseases with application to heart disease. *Amer. Heart J.,* **45,** 644

34. Le Compte, P. M. (1955). Vascular lesions in diabetes mellitus. *J. Chron. Dis.,* **2,** 178

35. Garcia, M. J., McNamara, P. M., Gordon, T. and Kannel, W. B. (1974). Morbidity and mortality in diabetics in the Framingham population. Sixteen year follow-up study. *Diabetes,* **23,** 105

36. Ostrander, L. D., Francis, T., Hayner, N. S., Kjelsberg, M. O. and Epstein, F. H. (1965). The relationship of cardiovascular disease to hyperglycemia. *Ann. Int. Med.,* **62,** 1188

37. Epstein, F. H. (1967). Some uses of prospective observations in the Tecumseh community health study. *Proc. Roy. Soc. Med.,* **60,** 56

38. Keen, H., Rose, G., Pyke, D. A., Boyns, D., Chlouverakis, C. and Mistry, S. (1965). Blood sugar and arterial disease. *Lancet,* **ii,** 505

39. Keen, H. and Jarrett, R. J. (1973). Macroangiopathy — its prevalence in asymptomatic diabetes. *Adv. Metab. Disorders* **(Suppl. 2),** 3

40. Pell, S. and D'Alonzo, C. A. (1970). Factors associated with long-term survival of diabetes. *J. Amer. Med. Ass.,* **214,** 1833

41. Goldstein, J. L., Hazzard, W. R., Schrott, H. G., Bierman, E. L. and Motulsky, A. G. (1973). Hyperlipidemia in coronary heart disease. Lipid levels in 500 survivors of myocardial infarction. *J. Clin. Invest.,* **52,** 1533

42. Robertson, W. B. and Strong, J. P. (1968). Atherosclerosis in persons with hypertension and diabetes mellitus. *Lab. Invest.,* **18,** 538

43. Ricketts, H. T. (1955). The problem of degenerative vascular disease in diabetes. *Amer. J. Med.,* **19,** 933

44. Bradley, R. F. and Bryfogle, J. W. (1956). Survival of diabetic patients after myocardial infarction. *Amer. J. Med.,* **20,** 207

45. Downie, E. and Martin, F. I. R. (1959). Vascular disease in juvenile diabetic patients of long duration. *Diabetes,* **8,** 383

46. New, M. I., Roberts, T. N., Bierman, E. L. and Reader, G. G. (1963). The significance of blood lipid alterations in diabetes mellitus. *Diabetes,* **12,** 208

47. Reinheimer, W., Bliffen, G., McCoy, J., Wallace, D. and Albrink, M. J. (1967). Weight gain, serum lipids and vascular disease in diabetics. *Amer. J. Clin. Nutr.,* **20,** 986

48. University Group Diabetes Program (1970). A study of the effects of hypoglycemic agents on the vascular complications of patients with adult-onset diabetes. *Diabetes,* **19 (Suppl. 2)**

49. Schor, S. S. (1973). Statistical problems in clinical trials. The UGDP study revisited. *Amer. J. Med.,* **55,** 727

50. Keen, H., Jarrett, R. J., Ward, J. D. and Fuller, J. H. (1973). Borderline diabetics and their response to tolbutamide. *Adv. Metab. Disorders,* **(Suppl. 2),** 521

51. Passikivi, J. (1973). Long-term treatment of patients with abnormal intravenous glucose tolerance after myocardial infarction. *Adv. Metab. Disorders,* **(Suppl. 2),** 533

52. Tzagournis, M. and Reynertson, R. (1972). Mortality from coronary heart disease during phenformin therapy. *Ann. Int. Med.,* **76,** 587

53. Carlson, L. A. and Bottiger, L. E. (1972). Ischaemic heart disease in relation to fasting values of plasma triglycerides and cholesterol. *Lancet,* **i,** 865

54. Gossain, V. V. and Ahuja, M. M. S. (1967). Dietetic analysis and blood lipids, chemical and isotopic studies in vascular disease among Indian diabetics. *Amer. J. Clin. Nutr.,* **20,** 834

55. Santen, R. J., Willis, P. W. and Fajans, S. S. (1972). Atherosclerosis in diabetes mellitus. *Arch. Int. Med.,* **130,** 833

56. Ostrander, L. D., Neff, B. J., Block, W. D., Francis, T. and Epstein, F. H. (1967). Hyperglycemia and hypertriglyceridemia among persons with coronary heart disease. *Ann. Int. Med.,* **67,** 34

57. Lowy, A. D. and Barach, J. H. (1958). Predictive value of lipoprotein and cholesterol determinations in diabetic patients who develop cardiovascular complications. *Circulation,* **17,** 14

58. Society of Actuaries (1959). *Build and Blood Pressure Study, 1959* (Chicago: Society of Actuaries)

59. Keys, A., Aravanis, C., Blackburn, H., Van Buchem, F. S. P., Buzino, R., Djordjevic, B. S., Fidanza, F., Karvonen, M. J., Menotli, A., Puddo, V. and Taylor, H. L. (1972). Coronary heart disease: overweight and obesity as risk factors. *Ann. Int. Med.,* **77,** 15

60. Bagdade, J. D., Bierman, E. L. and Porte, D., Jr (1971). Influence of obesity on the relationship between insulin and triglyceride levels in endogenous hypertriglyceridaemia. *Diabetes,* **20,** 664

61. Montoye, H. J., Epstein, F. H. and Kjelsberg, M. O. (1966). Relationship between serum cholesterol and body fatness, an epidemiological study. *Amer. J. Clin. Nutr.,* **18,** 397

62. Kannel, W. B., Brand, N., Skinner, J. J., Dawber, T. R. and McNamara, P. M. (1967). The relation of adiposity to blood pressure and development of hypertension. The Framingham study. *Ann. Int. Med.,* **67,** 48

63. Pyke, D. A. and Please, N. W. (1957). Obesity, parity and diabetes. *J. Endocrinol.,* **15,** xxvi

64. Nestel, P. J., Whyte, H. M. and Goodman, D. S. (1969). Distribution and turnover of cholesterol in humans. *J. Clin. Invest.,* **48,** 982

65. Miettinen, T. A. (1971). Cholesterol production in obesity. *Circulation,* **44,** 842

66. Chiang, B. N., Perlman, L. V. and Epstein, F. H. (1969). Overweight and hypertension. *Circulation,* **39,** 403
67. Veterans Administration Co-operative Study Group on Antihypertensive Agents (1967). Effects of treatment on morbidity in hypertension. *J. Amer. Med. Ass.,* **202,** 1028
68. Veterans Administration Co-operative Study Group on Antihypertensive Agents (1970). Effects of treatment on morbidity in hypertension. II. Results in patients with diastolic blood pressure averaging 90 through 114 mm Hg. *J. Amer. Med. Ass.,* **213,** 1143
69. Kannel, W. B., Castelli, W. P., McNamara, P. M., McKee, P. A. and Feinleib, M. (1972). Role of blood pressure in the development of congestive heart failure. The Framingham study. *New Engl. J. Med.,* **287,** 781
70. Goldenberg, S., Alex, M. and Blumenthal, H. T. (1958). Sequelae of arteriosclerosis of the aorta and coronary arteries. *Diabetes,* **7,** 98
71. Pell, S. and D'Alonzo, C. A. (1967). Some aspects of hypertension in diabetes mellitus. *J. Amer. Med. Ass.,* **202,** 10
72. Epstein, F. H. (1967). Hyperglycemia. A risk factor in coronary heart disease. *Circulation,* **36,** 609
73. Pyke, D. A. (1968). Arterial disease and diabetes. In: *Clinical Diabetes and its Biochemical Basis,* 506 (W. G. Oakley, D. A. Pyke, and K. W. Taylor, editors) (Oxford: Blackwell)
74. Freedman, P., Moulton, R. and Spencer, A. G. (1958). Hypertension and diabetes mellitus. *Quart. J. Med., N. S.,* **27,** 293
75. Karlefors, T. (1966). Haemodynamic studies in male diabetics. *Acta Med. Scand.,* **(Suppl. 449),** 45
76. Albrink, M. J., Lavietes, P. H., and Man, E. B. (1963). Vascular disease and serum lipids in diabetes mellitus. *Ann. Int. Med.,* **58,** 305
77. Ahrens, E. H., Hirsch, J., Oette, K., Farquhar, J. W. and Stein, Y. (1961). Carbohydrate-induced and fat-induced lipemia. *Trans. Ass. Amer. Phys.,* **74,** 134
78. Bierman, E. L. and Hamlin, J. T. (1961). The hyperlipemic effect of a low-fat, high-carbohydrate diet in diabetic subjects. *Diabetes,* **10,** 432
79. Stone, D. B. and Connor, W. E. (1963). The prolonged effects of a low cholesterol, high carbohydrate diet upon the serum lipids in diabetic patients. *Diabetes,* **12,** 127
80. Antonis, A. and Bersohn, I. (1961). The influence of diet on serum triglycerides. *Lancet,* **i,** 3
81. Schlierf, G., Reinheimer, W. and Stossberg, V. (1971). Diurnal patterns of plasma triglycerides and free fatty acids in normal subjects and patients with endogenous (type IV) hyperlipoproteinemia. *Nutr. Metab.,* **13,** 80
82. Brunzell, J. D., Lerner, R. L., Hazzard, W. R., Porte, D., Jr and Bierman, E. L. (1971). Improved glucose tolerance with high carbohydrate feeding in mild diabetes. *New Engl. J. Med.,* **284,** 521
83. Brunzell, J. D., Lerner, R. L., Porte, D., Jr and Bierman, E. L. (1974). Effect of a fat-free, high carbohydrate diet on diabetic subjects with fasting hyperglycemia. *Diabetes,* **23,** 138
84. Porte, D., Jr and Bagdade, J. D. (1970). Human insulin secretion: an integrated approach. *Ann. Rev. Med.,* **21,** 219
85. Bagdade, J. D., Bierman, E. L. and Porte, D., Jr (1967). The significance of basal insulin levels in evaluation of the insulin response to glucose in diabetic and non-diabetic subjects. *J. Clin. Invest.,* **46,** 1549

86. Jackson, W. P. U., van Mieghem, W. and Keller, P. (1972). Insulin excess as the initial lesion in diabetes. *Lancet*, **i**, 1040

87. Reaven, G. M., Olefsky, J. and Farquhar, J. W. (1972). Does hyperglycaemia or hyperinsulinaemia characterise the patient with chemical diabetes. *Lancet*, **i**, 1247

88. Grodsky, G. M. (1965). Production of autoantibodies to insulin in man and rabbits. *Diabetes*, **14**, 396

89. Pearson, M. J. and Martin, F. I. R. (1970). The separation of total plasma insulin from binding proteins using gel filtration: its application to the measurement of rate of insulin disappearance. *Diabetologia*, **6**, 581

90. Heding, L. G. (1972). Determination of total serum insulin (IRI) in insulin-treated diabetic patients. *Diabetologia*, **8**, 260

91 Nakagawa, S., Nakayama, H., Sasaki, T., Yoshino, K., Yu, Y. Y., Shinosaki, K., Aoki, S. and Mashimo, K. (1973). A simple method for the determination of serum free insulin levels in insulin-treated patients. *Diabetes*, **22**, 590

92. Peters, N. and Hales, C. N. (1965). Plasma insulin concentration after myocardial infarction. *Lancet*, **i**, 1144

93. Nikkila, E. A., Miettinen, T. A., Vesenne, M. R. and Pelkonen, R. (1965). Plasma insulin in coronary heart disease. *Lancet*, **ii**, 508

94. Tzagournis, M., Chiles, R., Ryan, J. M. and Skillman, T. G. (1968). Interrelationships of hyperinsulinism and hypertriglyceridemia in young patients in coronary heart disease. *Circulation*, **38**, 1156

95. Sloan, J. M., Mackay, J. S. and Sheridan, B. (1971). The incidence of plasma insulin, blood sugar and serum lipid abnormalities in patients with atherosclerotic disease. *Diabetologia*, **7**, 431

96. Duff, G. L. and McMillan, G. C. (1949). The effect of alloxan diabetes on experimental atherosclerosis in the rabbit. I. The inhibition of experimental atherosclerosis in alloxan diabetes. II. The effect of alloxan diabetes on the retrogression of experimental cholesterol atherosclerosis. *J. Exp. Med.*, **89**, 611

97. McGill, H. G. and Holman, R. L. (1949). The influence of alloxan diabetes on cholesterol atheromatosis in the rabbit. *Proc. Soc. Exp. Biol. Med.*, **72**, 72

98. Duff, G. L., Brechin, J. H. and Finkelstein, W. E. (1954). The effect of alloxan diabetes on experimental cholesterol atherosclerosis in the rabbit. IV. The effect of insulin therapy on the inhibition of atherosclerosis in the alloxan-diabetic rabbit. *J. Exp. Med.*, **100**, 371

99. Cook, D. L., Mills, L. M. and Green, O. M. (1954). The mechanism of alloxan protection in experimental atherosclerosis. *J. Exp. Med.*, **99**, 119

100. Beveridge, J. M. R. and Johnson, S. E. (1950). Studies on diabetic rats: the production of cardiovascular and renal disease in diabetic rats. *Brit. J. Exp. Pathol.*, **31**, 285

101 Lehner, N. D. M., Clarkson, T. B. and Lofland, H. B. (1971). The effect of insulin deficiency, hypothyroidism and hypertension on atherosclerosis in the squirrel monkey. *Exp. Molec. Pathol.*, **15**, 230

102. Kalant, N., Teitelbaum, J. I., Cooperberg, A. A. and Harland, W. A. (1964). Dietary atherogenesis in alloxan diabetes. *J. Lab. Clin. Med.*, **63**, 147

103. Pierce, F. T. (1952). The relationship of serum lipoproteins to atherosclerosis in the cholesterol-fed alloxanized rabbit. *Circulation*, **5**, 401

104. Kalant, N. and Harland, W. A. (1961). The effect of an atherogenic diet on normal and alloxan-diabetic rats. *Can. Med. Ass. J.*, **84**, 251

105. Wilson, R. B., Martin, J. M. and Hartroft, W.S. (1967). Evaluation of the relative pathogenic roles of diabetes and serum cholesterol levels in the development of cardio-

vascular lesions in rats. *Diabetes,* **16,** 71

106. Matsuda, I. and Kalant, M. (1966). Effect of alloxan diabetes on cholesterol flux into aorta. *Diabetes,* **15,** 604
107. Wilson, R. B., Martin, J. M. and Hartroft, W. S. (1969). Failure of insulin therapy to prevent cardiovascular lesions in diabetic rats fed an atherogenic diet. *Diabetes,* **18,** 225
108. Cruz, A. B., Amatuzio, D. S., Grande, F. and Hay, L. J. (1961). Effect of intra-arterial insulin on tissue cholesterol and fatty acids in alloxan-diabetic dogs. *Circulation Res.,* **9,** 39
109. Foster, D. W. and Siperstein, M. D. (1960). Effect of diabetes on cholesterol and fatty acid synthesis in the rat aorta. *Amer. J. Physiol.,* **198,** 25
110. Wertheimer, H. E. and Bentor, V. (1961). Physiologic and pathologic influences on the metabolism of rat aorta. *Circulation Res.,* **9,** 23
111. Mulcahy, P. D. and Winegrad, A. I. (1962). Effects of insulin and alloxan diabetes on glucose metabolism in rabbit aortic tissue. *Amer. J. Physiol.,* **203,** 1038
112. Urrutia, G., Beavan, D. W. and Cahill, G. F. (1962). Metabolism of glucose-U-C^{14} in rat aorta *in vitro. Metabolism,* **11,** 530
113. Mahler, R. F. (1965). Diabetes and arterial lipids. *Quart. J. Med., N.S.,* **34,** 484
114. McGill, H.C., Jr(1968). Fatty streaks in coronary arteries and aorta. *Lab. Invest.,* **18,** 560
115. Parker, F. (1960). An electron microscope study of experimental atherosclerosis. *Amer. J. Pathol.,* **36,** 19
116. Geer, J. C., McGill, H. C. and Strong, J. P. (1961). The fine structure of human atherosclerotic lesions. *Amer. J. Pathol.,* **38,** 263
117. Haust, M. D. and More, R. H. (1963). Significance of the smooth muscle cell in atherogenesis. In: *Evolution of the Atherosclerotic Plaque,* 51 (R. J. Jones, editor) (Chicago: University of Chicago Press)
118. Daoud, A., Jarmolych, J., Zumbo, A. and Fani, K. (1964).'Preatheroma' phase of coronary atherosclerosis in man. *Exp. Molec. Pathol.,* **3,** 475
119. Benditt, E. P. and Benditt, J. M. (1973). Evidence for a monoclonal origin of human atherosclerotic plaques. *Proc. Nat. Acad. Sci., U.S.A.,* **70,** 1753
120. Ross, R. and Glomset, J. A. (1973). Atherosclerosis and the smooth muscle cell. *Science,* **180,** 1332
121. Adams, C. W. M. (1973). Tissue changes and lipid entry in developing atheroma. In: *Atherogenesis: Initiating Factors,* 5 [Ciba Foundation Symposium 12 (new series)]
122. Whereat, A. F. (1967). Atherosclerosis and metabolic disorder of the arterial wall. *Exp. Molec. Pathol.,* **7,** 233
123. Dayton, S. and Hashimoto, S. (1970). Cholesterol flux and metabolism in arterial tissue and atheromata. *Exp. Molec. Pathol.,* **13,** 253
124. Newman, H. A. and Zilversmit, D. B. (1959). Origin of various lipids in atheromatous lesions of rabbits. *Circulation,* **20,** 967
125. Smith, E. B. and Slater, R. S. (1973). Lipids and low-density lipoproteins in intima in relation to its morphological characteristics. In: *Atherogenesis: Initiating Factors,* 39 (Ciba Foundation Symposium 12 (new series)
126. Stein, O., Stein, Y. and Eisenberg, S. (1973). A radioautographic study of the transport of ^{125}I-labelled serum lipoproteins in rat aorta. *Z. Zellforsch.,* **138,** 223
127. Stein, Y. and Stein, O. (1973). Lipid synthesis and degradation and lipoprotein transport in mammalian aorta. In: *Atherogenesis: Initiating Factors,* 165 [Ciba Foundation Symposium 12 (new series)]

128. Chobanian, A. V. and Manzur, F. (1972). Metabolism of lipid in the human fatty streak lesion. *J. Lipid Res.*, **13**, 201

129. Siperstein, M. D., Chaikoff, I. L. and Chernick, S. S. (1951). Significance of endogenous cholesterol in arteriosclerosis: synthesis in arterial tissue. *Science*, **113**, 747

130. Dayton, S., Hashimoto, S. and Jessaney, J. (1961). Cholesterol kinetics in the normal rat aorta and the influence of different types of dietary fat. *J. Atheroscler. Res.*, **1**, 444

131. Chobanian, A. V. (1968). Sterol synthesis in the human arterial intima. *J. Clin. Invest.*, **47**, 595

132. Wahlqvist, M. L., Day, A. J. and Tume, R. K. (1969). Incorporation of oleic acid into lipid by foam cells in human atherosclerotic lesions. *Circulation Res.*, **24**, 123

133. Stein, Y. and Stein, O. (1962). Incorporation of fatty acids into lipids of aortic slices of rabbits, dogs, rats and baboons. *J. Atheroscler. Res.*, **2**, 400

134. Parker, F., Schimmelbusch, W. and Williams, R. H. (1964). The enzymatic nature of phospholipid synthesis in the normal rabbit and human aorta. *Diabetes*, **13**, 182

135. Eisenberg, S., Stein, Y. and Stein, O. (1969). Phospholipases in arterial tissue. IV. *J. Clin. Invest.*, **48**, 2320

136. Stein Y., Stein, O. and Shapiro, B. (1963). Enzymic pathways of glyceride and phospholipid synthesis in aortic homogenates. *Biochim. Biophys. Acta*, **70**, 33

137. Stout, R. W. (1971). The effect of insulin on the incorporation of (I-^{14}C) sodium acetate into the lipids of the rat aorta. *Diabetologia*, **7**, 367

138. Ross, R. and Klebanoff, S. J. (1971). The smooth muscle cell. I. *In vivo* synthesis of connective tissue proteins. *J. Cell. Biol.*, **50**, 159

139. Ross, R. (1971). The smooth muscle cell. II. Growth of smooth muscle in culture and formation of elastic fibers. *J. Cell. Biol.*, **50**, 172

140. Ross, R., Glomset, J., Kariya, B. and Harker, L. H. (1974). A platelet-dependent serum factor that stimulates the proliferation of arterial smooth muscle cells *in vitro*. *Proc. Nat. Acad. Sci., U.S.A.*, **71**, 1207

141. Bierman, E. L., Stein, O. and Stein, Y. (1974). Lipoprotein uptake and metabolism by rat aortic smooth muscle cells in tissue culture. *Circulation Res*, **35**, 136

142. Kramsch, D. M. and Hollander, W. (1973). The interaction of serum and arterial lipoproteins with elastin of the arterial intima and its role in the lipid accumulation in atherosclerotic plaques. *J. Clin. Invest.*, **52**, 236

143. Bierman, E. L., Eisenberg, S., Stein, O. and Stein, Y. (1973). Very low density lipoprotein 'remnant' particles: uptake by aortic smooth muscle cells in culture. *Biochim. Biophys. Acta*, **329**, 163

144. Zilversmit, D. B. (1973). A proposal linking atherogenesis to the interaction of endothelial lipoprotein lipase with triglyceride-rich lipoproteins. *Circulation Res.*, **33**, 633

145. Gabbay, J. H. (1973). The sorbitol pathway and the complications of diabetes. *New Engl. J. Med.*, **288**, 831

146. Clements, R. S., Jr, Morrison, A. D. and Winegrad, A. I. (1969). Polyol pathway in aorta: regulation by hormones. *Science*, **166**, 1007

147. Morrison, A. D., Clements, R. S. and Winegrad, A. I. (1972). Effects of elevated glucose concentrations on the metabolism of the aortic wall. *J. Clin. Invest.*, **51**, 3114

148. Stout, R. W., Bierman, E. L. and Ross, R. (1975). The effect of insulin on the proliferation of cultured primate arterial smooth muscle cells. *Circulation Res.*, **36**, 319

149. Stout, R. W. (1968). Insulin-stimulated lipogenesis in arterial tissue in relation to

diabetes and atheroma. *Lancet,* **ii,** 702

150. Stout, R. W. (1969). Insulin stimulation of cholesterol synthesis by arterial tissue. *Lancet,* **ii,** 467

151. Stout, R. W., Buchanan, K. D. and Vallance-Owen, J. (1972). Arterial lipid metabolism in relation to blood glucose and plasma insulin in rats with streptozotocin-induced diabetes. *Diabetologia,* **8,** 398

152. Stamler, J., Pick, R. and Katz, L. N. (1960). Effect of insulin in the induction and regression of atherosclerosis in the chick. *Circulation Res.,* **8,** 572

153. Stout, R. W. (1970). Development of vascular lesions in insulin-treated animals fed a normal diet. *Brit. Med. J.,* **3,** 685

154. Stout, R. W., Buchanan, K. D. and Vallance-Owen, J. (1973). The relationship of arterial disease and glucagon metabolism in insulin-treated chickens. *Atherosclerosis,* **18,** 153

155. Renold, A. E., Gonet, A. E., Stauffacher, W. and Jeanrenaud, B. (1968). Laboratory animals with spontaneous diabetes and/or obesity; suggested suitability for the study of spontaneous atherosclerosis. *Prog. Biochem. Pharmacol.,* **4,** 363

156. Stout, R. W. and Vallance-Owen, J. (1969). Insulin and atheroma. *Lancet,* **i,** 1078

157. Oakley, W. G., Pyke, D. A., Tattersall, R. B. and Watkins, P. J. (1974). Long-term diabetes. *Quart. J. Med.,* **43,** 145

158. Wolinsky, H. (1973). Mesenchymal response of the blood vessel wall. A potential avenue for understanding and treating atherosclerosis. *Circulation Res.,* **32,** 543

159. Thannhauser, S. J. (1958). Hyperlipidemia in severe untreated diabetes with secondary eruptive xanthoma. In: *Lipidoses: Diseases of the Intracellular Lipid Metabolism, 3rd edition,* 296 (New York and London: Grune and Stratton)

160. Man, E. B. and Peters, J. P. (1935). Serum lipids in diabetes. *J. Clin. Invest.,* **14,** 579

161. Gofman, J. W., Delalla, O., Glazier, F., Freeman, N. K., Nichols, A. V., Strisower, B. and Tamplin, A. R. (1954). The serum lipoprotein transport system in health, metabolic disorders, atherosclerosis and coronary heart disease. *Plasma,* 2, 413

162. Bagdade, J. D., Porte, D., Jr and Bierman, E. L. (1967). Diabetic lipemia. A form of acquired fat-induced lipemia. *New Eng. J. Med.,* **276,** 427

163. Chance, G. W., Albutt, E. C. and Edkins, S. M. (1969). Serum lipids and lipoproteins in untreated diabetic children. *Lancet,* **i,** 1126

164. Wilson, D. E., Schreibman, P. H., Day, V. C. and Arky, R. A. (1970). Hyperlipidemia in an adult diabetic population. *J. Chron. Dis.,* 23, 501

165. Brunzell, J. D., Robertson, R. P., Lerner, R. L., Hazzard, W. R., Ensinck, J. W., Bierman, E. L. and Porte, D., Jr (1975). The relation of fasting glucose levels to insulin secretion during an intravenous glucose tolerance test (in preparation)

166. Belknap, B. H., Bagdade, J. D., Amaral, J. A. P. and Bierman, E. L. (1967). Plasma lipids and mild glucose intolerance. II. A double blind study of the effect of tolbutamide and placebo in mild adult diabetic outpatients. In: *Tolbutamide After Ten Years,* 171 (W. J. H. Butterfield and W. Van Westering, editors) (Amsterdam: Excerpta Medica)

167. Bierman, E. L. and Porte, D., Jr (1968). Carbohydrate intolerance and lipemia. *Ann. Int. Med.,* **68,** 926

168. Christensen, N. J. (1974). Plasma norepinephrine and epinephrine in untreated diabetes, during fasting and after insulin administration. *Diabetes,* **23,** 1

169. Balasse, E. O., Bier, D. M. and Havel, R. J. (1972). Early effects of anti-insulin serum on hepatic metabolism of plasma free fatty acids in dogs. *Diabetes,* **21,** 280

170. Woodside, W. F. and Heimberg, M. (1972). Hepatic metabolism of free fatty acids in experimental diabetes. *Israel J. Med. Sci.*, **8,** 309

171. Meissner, W. A. and Legg, M. A. (1971). The pathology of diabetes. In: *Joslin's Diabetes Mellitus, 11th edition,* 179 (A. Marble, P. White, R. F. Bradley and L. P. Krall, editors) (Philadelphia: Lea and Febiger)

172. Basso, L. V. and Havel, R. J. (1970). Hepatic metabolism of free fatty acids in normal and diabetic dogs. *J. Clin. Invest.*, **49,** 537

173. Reaven, E. P. and Reaven, G. M. (1974). Evidence for multiple causality in the development of hypertriglycerdemia in insulin-deficient rats. *Diabetes*, **23 (Suppl 1),** 346

174. Gross, R. C. and Carlson, L. A. (1968). Metabolic effects of nicotinic acid in acute insulin deficiency in the rat. *Diabetes*, **17,** 353

175. Porte, D. Jr (1969). Sympathetic regulation of insulin secretion and its relation to diabetes mellitus. *Arch. Int. Med.*, **123,** 252

176. Bolzano, K., Sandhofer, F., Sailer, S. and Braunsteiner, H. (1972). The effect of oral administration of sucrose on the turnover rate of plasma free fatty acids and on the esterification rate of plasma free fatty acids to plasma triglycerides in normal subjects, patients with primary endogenous hypertriglyceridemia and patients with well controlled diabetes mellitus. *Horm. Metab. Res.*, **4,** 446

177. Nikkilä, E. A. and Kekki, M. (1973). Plasma triglyceride transport kinetics in diabetes mellitus. *Metabolism*, **22,** 1

178. Nikkilä, E. A. and Kekki, M. (1971). Polymorphism of plasma triglyceride kinetics in normal human adult subjects. *Acta Med. Scand.*, **190,** 49

179. Nikkilä, E. A. and Kekki, M. (1972). Plasma endogenous triglyceride transport in hypertriglyceridemia and effect of a hypolipidaemic drug (SU-13437). *Eur. J. Clin. Invest.*, **2,** 231

180. Lewis, B., Mancini, M., Mattock, M., Chait, A. and Fraser, T. R. (1972). Plasma triglyceride and fatty acid metabolism in diabetes mellitus. *Eur. J. Clin. Invest.*, **2,** 445

181. Brunzell, J. D., Hazzard, W. R., Porte, D. Jr, and Bierman, E. L. (1973). Evidence for a common, saturable, triglyceride removal mechanism for chylomicrons and very low density lipoproteins in man. *J. Clin. Invest.*, **52,** 1578

182. Blanchette-Mackie, E. J. and Scow, R. O. (1971). Sites of lipoprotein lipase activity in adipose tissue perfused with chylomicrons. Electron microscope cytochemical study. *J. Cell. Biol.*, **51,** 1

183. Wilson, D. E., Schreibman, P. H. and Arky, R. A. (1969). Postheparin lipolytic activity in diabetic patients with a history of mixed hyperlipemia. Relative rates against artificial substrates and human chylomicrons. *Diabetes*, **18,** 562

184. Brunzell, J. D., Porte, D., Jr, and Bierman, E. L. (1975). Reversible abnormalities in post-heparin lipolytic activity and hypertriglyceridemia in diabetes (submitted for publication)

185. Aktin, E. and Meng, H. C. (1972). Release of clearing factor lipase (lipoprotein lipase) *in vivo* and from isolated perfused hearts of alloxan diabetic rats. *Diabetes*, **21,** 149

186. Schnatz, J. D. and Williams, R. H. (1963). The effect of acute insulin deficiency on adipose tissue lipolytic activity and plasma lipids. *Diabetes*, **12,** 174

187. Pykälistö, O. (1970). *Regulation of Adipose Tissue Lipoprotein Lipase by Free Fatty Acids* (Thesis, University of Helsinki, Finland)

188. Pykälistö, O. J., Smith, P. H., Bierman, E. L. and Brunzell, J. D. (1974). Decreased adipose tissue lipoprotein lipase in untreated diabetic man. *Diabetes,* **(Suppl. 1),** 348

Diabetes

189. Persson, B. (1973). Lipoprotein lipase activity of human adipose tissue in health and in some diseases with hyperlipidemia as a common feature. *Acta Med. Scand.*, **193**, 457

190. Bagdade, J. D., Porte, D.,Jr and Bierman, E. L. (1968). Acute insulin withdrawal and the regulation of plasma triglyceride removal in diabetic subjects. *Diabetes*, **17**, 127

191. Fredrickson, D. S. and Levy, R. I. (1972). In: *Metabolic Basis of Inherited Disease, 3rd edition*, 598 (J. B. Stanbury, J. B. Wyngaarden and D. S. Fredrickson, editors) (New York: McGraw-Hill)

192. Goldstein, J. L., Hazzard, W. R., Schrott, H. G., Bierman, E. L. and Motulsky, A. G. (1973). Hyperlipidemia in coronary heart disease. II. Genetic analysis of lipid levels in 176 families and delineation of a new inherited disorder: combined hyperlipidemia. *J. Clin. Invest.*, **52**, 1544

193. Brunzell, J. D., Hazzard, W. R., Motulsky, A. G. and Bierman, E. L. (1974). Prevalence of diabetes in hypertriglyceridemia. *Diabetes*, **23 (Suppl. I),** 351

194. Albrink, M. J. and Davidson, P. C. (1966). Impaired glucose tolerance in patients with hypertriglyceridaemia. *J. Lab. Clin. Med.*, **67**, 573

195. Bagdade, J. D., Porte, D., Jr and Bierman, E. L. (1968). Hypertriglyceridemia: a metabolic consequence of chronic renal failure. *New Engl. J. Med.*, **279**, 181

196. Robertson, R. P., Gavereski, D. J., Henderson, J. D., Porte, D.,Jr and Bierman, E. L.,(1973). Accelerated triglyceride secretion: a metabolic consequence of obesity. *J. Clin. Invest.*, **52**, 1620

197. Topping, D. L. and Mayes, P. A. (1972). The immediate effects of insulin and fructose on the metabolism of the perfused liver: changes in lipoprotein secretion, fatty acid oxidation and esterification, lipogenesis and carbohydrate metabolism. *Biochem. J.*, **126**, 295

198. Reaven, G. M., Lerner, R. L., Stern, M. P. and Farquhar, J. W. (1967). Role of insulin in endogenous hypertriglyceridemia. *J. Clin. Invest.*, **46**, 1756

199. Eaton, R. P. and Schade, D. S. (1973). Glucagon resistance as a hormonal basis for endogenous hyperlipaemia. *Lancet*, **i,** 973

200. Olefsky, J., Reaven, G. M. and Farquhar, J. W. (1974). Effects of weight reduction on obesity. Studies on lipid and carbohydrate metabolism in normal and hyperlipoproteinemic subjects. *J. Clin. Invest.*, **53**, 64

201. Davidson, P. C. and Albrink, M. J. (1965). Insulin resistance in hyperglyceridemia. *Metabolism*, **14**, 1059

202. Kekki, M. and Nikkilä, E. A. (1971). Plasma triglyceride metabolism in the adult nephrotic syndrome. *Eur. J. Clin. Invest.*, **1**, 345

203. Belknap, B. H., Amaral, J. A. P. and Bierman, E. L. (1967). Plasma lipids and mild glucose tolerance. I. The response of plasma triglycerides to high carbohydrate feeding and the effect of tolbutamide therapy. In: *Tolbutamide After Ten Years*, 159 (W. J. H. Butterfield and W. Van Westering, editors) (Amsterdam: Excerpta Medica)

204. Fredrickson, D. S. and Levy, R. I. (1972). In: *Metabolic Basis of Inherited Disease, 3rd edition*, 275 (J. B. Stanbury, J. B Wyngaarden and D. S. Fredrickson, editors) (New York: McGraw-Hill)

205. Schofield, G., Birge, C., Miller, J. P., Kessler, G. and Santiago, J. (1974). Apolipoprotein B levels and altered lipoprotein composition in diabetes. *Diabetes*, **23**, 827

206. Brunzell, J. D. and Schrott, H. G. (1973). The interaction of familial and secondary causes of hypertriglyceridemia: role in pancreatitis. *Trans. Ass. Amer. Physins.*, **86**, 245

207. Parker, F., Bagdade, J. D., Odland, G. F. and Bierman, E. L. (1970). Evidence for

the chylomicron origin of lipids accumulating in diabetic eruptive xanthomas: a correlative lipid biochemical, histochemical and electron microscopic study. *J. Clin. Invest.,* **49,** 2172

208.   Brunzell, J. D., Shankle, S. W. and Bethune, J. E. (1968). Congenital generalized lipodystrophy accompanied by cystic angiomastosis. *Ann. Int. Med.,* **69,** 501

209.   Havel, R. J., Basso, L. V. and Kane, J. P. (1967). Mobilization and storage of fat in congenital and late onset forms of total lipodystrophy. *Clin. Res.,* **15,** 48

6
SYNALBUMIN INSULIN ANTAGONISM

J. Vallance-Owen and J. S. Bajaj

INTRODUCTION

The first definitive evidence that diabetes mellitus could be associated with the pancreas stemmed from the studies of von Mering and Minkowski[1] who, in 1899, showed that totally depancreatised dogs developed hyperglycaemia and glycosuria, eventually dying in ketosis and coma. Then in 1921 Banting and Best[2] extracted insulin from the pancreas and found that this substance cured the carbohydrate intolerance of diabetics. This monumental discovery revolutionised the treatment of diabetes, but at the same time it appears to have retarded the research and the thinking about the true nature of diabetes in man for more than a quarter of a century. Not only was it generally considered thereafter that diabetes was due to a relative or absolute lack of insulin from a hereditary defective pancreas, but also much of the experimental work relating to the metabolic defect in diabetes was carried out in depancreatised and later alloxan-diabetic animals, the tacit assumption being that this situation was the same as diabetes mellitus seen in man.

There is evidence which is presented elsewhere in this volume that there may be abnormalities of formation and release of insulin in essential diabetics, and certainly many continue to believe that in some way diabetes is due to a primary failure of the pancreatic β-cells [3-5].

However, during the last thirty years, three important points have been established:

(1) Many diabetics need insulin in quantities far in excess of that needed by totally depancreatised man[6-8];

171

(2)   the morphological alternation found in the pancreas of diabetics is often quite inadequate to explain the biochemical findings[9-12];

(3)   biological and immunological assays of plasma insulin in diabetes have shown that the hormone is often present in normal or even increased amounts[13-17].

These facts are consistent with the theory that the primary fault may not lie in the pancreas but may be a result of the presence at the periphery of an anti-insulin factor. This situation can be investigated *in vitro* by using the rat diaphragm model introduced by Gemmill in 1941[18].

FUNDAMENTAL STUDIES

Small amounts of insulin accelerate both the utilisation of glucose and the synthesis of glycogen by the isolated rat diaphragm[18]. In addition, a quantitative relationship has been shown to exist between the concentration of insulin in the incubation medium and its effects on the glucose metabolism of the diaphragm[19 20].

By rigorous control of the experimental conditions and the employment of undiluted plasma this method can be used for the estimation of plasma insulin activity or the effective insulin concentration in undiluted plasma, i.e. the sum of the biological activity of insulin and its synergists on the one hand and its antagonists on the other[21 22].

There are two broad clinical groups of patients suffering from essential diabetes: those who require insulin treatment, without which they rapidly lose weight and become ketotic; and those who do not ordinarily need insulin and who show no tendency to develop ketosis unless their mild diabetes is complicated by infection. The latter are usually obese and often recover from the diabetic state when they are treated with a low carbohydrate, low calorie diet. Biological insulin activity was found in the plasma of all these mild, obese diabetics. In the fasting state, the values were usually a little above those found in a series of normal donors[13 23].

Release of insulin from liver/pancreas was demonstrated by monitoring the insulin response to oral glucose. A considerable rise in plasma insulin activity was found in samples taken from normal donors and obese diabetics one hour after ingestion of 50 g of glucose. Furthermore, when insulin was added *in vitro* to plasma from these two types its full activity was recovered, thus demonstrating the absence of measurable antagonism.

On the other hand, no measurable insulin activity was found in the plasma of untreated or uncontrolled essential diabetics who require insulin therapy to prevent ketosis, even though substantial doses of insulin, e.g. 70

units, had been injected one hour before the blood was withdrawn for testing. Moreover, when insulin was added to such plasma *in vitro*, the activity of the added insulin was markedly diminished.   When the patients within this group were controlled with sufficient insulin, or given insulin for the first time, to bring their blood sugar levels within the physiological range at the time of the test, essentially normal levels of plasma insulin activity were found and added insulin was no longer inhibited[13]. These initial observations were confirmed by Gundersen and Williams[24] who studied in particular untreated diabetics who subsequently needed insulin for survival.

These early studies on the insulin-requiring diabetics suggested that these patients need insulin to overcome some antagonist circulating in their plasma but which is not an antibody, and that in order to achieve control the quantity of insulin given must overcome this antagonist, and yield sufficient active or 'effective' insulin to carry out its normal physiological and metabolic functions.

INSULIN ANTAGONISM OF PLASMA ALBUMIN

The rat diaphragm model can also be used to estimate the effect of various plasma or serum fractions on the ability of insulin to promote glucose uptake by the tissue. Vallance-Owen, Dennes and Campbell observed that when the constituents of plasma were separated into various fractions there was an insulin antagonist associated with the protein albumin and which they subsequently termed synalbumin[25]. In concentrations of 3.5–5.5%, which included the physiological range in plasma, both diabetic and normal albumin completely inhibited the effect of 1000 μu/ml added insulin. At 1.25% however, the diabetic albumin was still highly antagonistic (synalbumin positive), whereas normal albumin was then inactive (synalbumin negative).

In whole plasma from normal subjects the antagonistic effect of the contained albumin appears to be completely masked by the production of adequate amounts of insulin from the pancreas. Thus, although the whole plasma from obese diabetics also exhibits no antagonism to insulin, it was of interest to study the antagonistic activity of the separated albumin fraction. At the same time albumin was isolated from the plasma of latent or prediabetics. 'Predibetes' has been defined by Jackson [26] as, 'the state of a person during the period before he or she becomes plainly and clinically diabetic, in which, however, there is a latent abnormality which may show itself under certain specific conditions'.

At a concentration of 1.25%, the albumin isolated from both the obese diabetics and the prediabetics was antagonistic to insulin to the same extent as albumin from the plasma of insulin-requiring diabetics, and to a considerably greater extent than normal albumin, which is inactive at this concentration[27]. It is also significant that patients with the diabetic syndrome of carbohydrate intolerance, as currently defined, but suffering from definite pancreatic disease, such as acute pancreatitis or haemochromatosis, or who have undergone total pancreatectomy, have no increased antagonism to insulin associated with their plasma albumin[28] [29].

These observations on the activity of the insulin antagonist associated with plasma albumin are summarised in Table 6.1.

Table 6.1 Antagonism to insulin of plamsa albumin, tested at 1.25% from normal subjects, essential and 'pancreatic' diabetic patients

Source of albumin tested at 1.25% and (no).	Mean glucose uptake above basal level.* ± SEM (mg glucose/100 ml/10 mg rat diaphragm)	
	buffer + 1000 μu insulin/ml	albumin in buffer + 1000 μu insulin/ml
Normal subjects (12)	13.91+0.68	13.24+0.86
Essential diabetics:		
insulin-requiring (12)	14.02+0.58	4.29+0.64
obese (13)	14.66+0.33	6.03+0.74
prediabetic (11)	15.04+0.54	6.23+0.67
'Pancreatic'** diabetics (10)	14.80+0.75	14.78+0.82

* Amount of glucose taken up by the diaphragm when no insulin is present in the incubation medium

** Following acute pancreatitis, haemochromatosis and pancreatectomy

Comparison of the activity found in normal donors with that found in the various diabetic groups indicates that a fundamental abnormality in essential or idiopathic diabetes mellitus is increased synalbumin antagonism to insulin action on muscle. This suggestion has received confirmation and considerable support from many quarters[30-43] but opposition from others[44-46]. Accordingly, controversy has built up and even now, sixteen years after the original demonstration of synalbumin, there are reports of artefactual insulin antagonists. Recently Holcomb and Dulin reported that albumin isolated by the trichloroacetic acid TCA method[47] contained an artefactual insulin antagonist but their data did not support the conclusion that this substance was derived from the dialysis membrane; they considered it was due to the TCA or to TCA protein complexes.

The first method which was used for the isolation of plasma albumin was a modification of the procedure published in 1957 by Debro *et al*[48]. This method involves the use of trichloroacetic acid to precipitate all proteins from the plasma sample followed by the treatment of the protein pellet with 1% TCA in ethanol to redissolve the albumin. To remove the solvents extensive dialysis of the extract is carried out using the semi-permeable membrane, Visking. Unfortunately there is a factor associated with the Visking which we showed is capable of antagonising insulin[49]. However, it was also shown that this artefactual antagonist does not contribute to the synalbumin effect on insulin[49].

Although other somewhat different methods of isolation of albumin have been tried, i.e. Cohn fractionation[25 31 34 36], either precipitation[25] and acid/methanol extraction[38] and found to yield synalbumin[25 31 34 36 38], to date the quickest method for the isolation of albumin is one which was adopted from the method of Fernandez *et al.* [50] This technique employs a non-aqueous ethanolic solution of hydrochloric acid for the separation of globulins from serum, followed by recovery of the albumin from the supernatant by precipitation with sodium acetate. For synalbumin tests the albumin pellet is then washed and dried with methanol, methanol ether and ether[51]. The results obtained with this type of isolate entirely endorse the previous results found by Vallance-Owen and co-workers using TCA albumin[51]. Furthermore, synalbumin extracts prepared at the same time from the same subjects by these two methods gave identical results[52].

Since insulin for use in physiological assays is normally dissolved in either hydrochloric or acetic acid and since this method uses only hydrochloric acid and sodium acetate, it is unlikely that albumin-bound anions from the reagent[47] are responsible for the antagonistic activity.

An interesting finding with albumin isolated by the modified Fernandez method was the level of antagonism gradually diminished on storage and after two weeks, had disappeared. Moreover once the antagonism was lost an insulin-like effect was found related to the previous degree of antagonism[51]. Subsequent studies have indicated that the antagonism can be conserved if the samples are stored in an oxygen-depleted atmosphere[53]. Parallel studies with TCA albumin have now been completed[54], which revealed that albumin isolated by the now obsolete TCA method is also vulnerable when stored in air and quickly loses its synalbumin activity. It is obvious therefore that many of the discordant results which have been attributed by some authors to differences in technique have been due to mismanagement of the extracts, there being a delay between extraction and testing with storage of the samples in air. This is particularly true in relation to the finding of insulin-like activity[44-46].

It has been shown now that fenfluramine significantly increases glucose uptake by rat diaphragm but this effect is not inhibited by synalbumin[55], indicating that it is only the insulin-promoted glucose uptake which is affected by synalbumin antagonism.

OTHER ASPECTS OF SYNALBUMIN ACTIVITY

Although synalbumin antagonises insulin action on muscle, it has also been repeatedly shown that it does not oppose the action of insulin on fat tissue[31] [37] [40] and in some hands, not only is the action of insulin unopposed, but there is an insulin-stimulatory effect as well[31] [40]. The importance of this observation will be stressed again in relation to obesity and ischaemic heart disease (see page 191). Moreover, in this connection synalbumin does have an anti-insulin effect on evoked responses from the feeding centre of the hypothalamus[56]. It has been shown that synalbumin prepared from diabetic patients significantly antagonises insulin-promoted glycine uptake and protein synthesis by the rat diaphragm[57] and these findings have now been confirmed[58]. Also it has been shown that the release of glucose into the incubation medium of liver homogenates from normal untreated rats increases with time and that previous intravenous insulin administration prevents this increase[59]. However, if insulin is injected together with albumin prepared from diabetic patients this effect of insulin is antagonised[59].

HORMONAL RELATIONSHIPS/DEPENDENCE

When albumin prepared from the plasma of insulin-requiring diabetics or normal subjects was passed through a partially acetylated cellulose column it lost its antagonistic activity to insulin, although still remaining electrophoretically identical with albumin. It was also noted that the plasma-albumin fraction from hypophysectomised subjects, whether initially diabetic or normal, was devoid of insulin antagonism[60]. Taken together, these lines of evidence indicate that the anti-insulin activity of the albumin fraction from normal subjects and diabetic patients is not due to the albumin itself but to some substance associated with it—hence the term, synalbumin antagonist. The results of hypophysectomy indicate that the antagonistic activity is related in some way to the pituitary gland; it has now been shown that the synalbumin antagonist is also dependent upon the adrenal corticosteroids[61]. These findings regarding hormone dependence are similar to those previously described for the insulin antagonism in the plasma of alloxan-diabetic rats[62] and depancreatised cats[63].

In view of the above, it was of interest that synalbumin antagonism changes during oral and intravenous glucose tolerance tests, the antagonism being inversely proportional to the glucose levels in the blood[64]. In these studies the blood sugar rose initially and more recently we have studied synalbumin during intravenous tolbutamide tolerance tests, where the immediate response is a fall in glucose levels[65]. These observations again showed a change in synalbumin related inversely to the blood sugar level and it is teleologically of interest that this should be so, bearing in mind the accepted actions of insulin in maintaining glucose homeostasis. This matter will be referred to again in relation to the nature of the synalbumin antagonist and its formation in the liver. The fundamental points relating to synalbumin are given in Table 6.2.

Table 6.2 Fundamental observations on synalbumin

(1) Human albumin antagonises insulin action on muscle

(2) It does not antagonise insulin action on adipose tissue

(3) Albumin from diabetic patients, including those in the potential or 'prediabetic' phase, is more antagonistic than albumin from normal subjects

(4) The antagonism depends upon the pituitary gland and the adrenal cortical steroids: it is not due to the albumin itself but to some substance associated with it — hence 'synalbumin'

(5) Synalbumin antagonism varies inversely with the blood sugar level in normal subjects

INHERITANCE OF DIABETES

It has long been recognised that there is a higher incidence of diabetes mellitus among the relatives of diabetics than among the rest of the population. Nevertheless, there is apparently considerable confusion regarding the mode of inheritance of this condition. After reviewing much of the available evidence, Pincus and White[66] and Steinberg[67] [68] conclude that predisposition to diabetes is probably due to homozygosity for a recessive gene; although a number of investigators hold different views or feel that this suggestion cannot explain all the facts, remembering that the studies to date have only involved carbohydrate intolerance. Circumstantial evidence concerning the behaviour of the syndrome of carbohydrate intolerance suggests that, if inheritance is involved, it is the liability to develop carbohydrate intolerance which is conferred by the appropriate genotype, and thus it is not possible to study the hereditary behaviour of essential diabetes mellitus without a suitable genetic marker[69].

The observations relating to synalbumin insulin antagonism outlined above do indicate that excessive synalbumin antagonism—the state of being synalbumin positive—can indeed be regarded as a suitable biochemical marker to ascertain whether or not a given person is constituted as an essential diabetic without reference to the symptom of carbohydrate intolerance. Accordingly, the genetic transmission of essential diabetes was studied by preparing albumin from the relatives of patients suffering from this condition. The result may be summarised by stating that excessive synalbumin antagonism is inherited as an autosomal 'dominant' characteristic[70]. Ninety-seven members of nine families have been studied. In the available sibships discounting any propositi, the ratio of synalbumin positive to synalbumin negative members did not differ significantly from 50:50. Overall, 39 were synalbumin negative and 58 were synalbumin positive, but only 16 of the latter had overt carbohydrate intolerance; a further three had recurrent spontaneous hypoglycaemia, while the remainder at the time of testing were quite asymptomatic, although in many cases overweight. These studies indicate that overt carbohydrate intolerance is relatively uncommon, or will be a late or very late event in the life of many people constituted as essential diabetics; they also imply that many more people are so constituted than was previously realised. Our own findings continue to indicate that approximately 25% of the general population in northern England[70] and in Northern Ireland[71] may be constituted as essential diabetics, with 18% so constituted in northern India[72 73]. An example of synalbumin inheritance is shown in Figure 6.1

Moreover, the insulin antagonism status of 303 individual albumin samples has a bimodal distribution[70] and a much smaller series of 142 random samples taken in a recent Belfast survey has a similar distribution[71].

A number of studies were also made on the albumin prepared from cord blood of synalbumin negative mothers and of synalbumin positive mothers who were either in the latent or overt phase of diabetes. Twenty-seven observations were made and the results are shown in Table 6.3. It can be

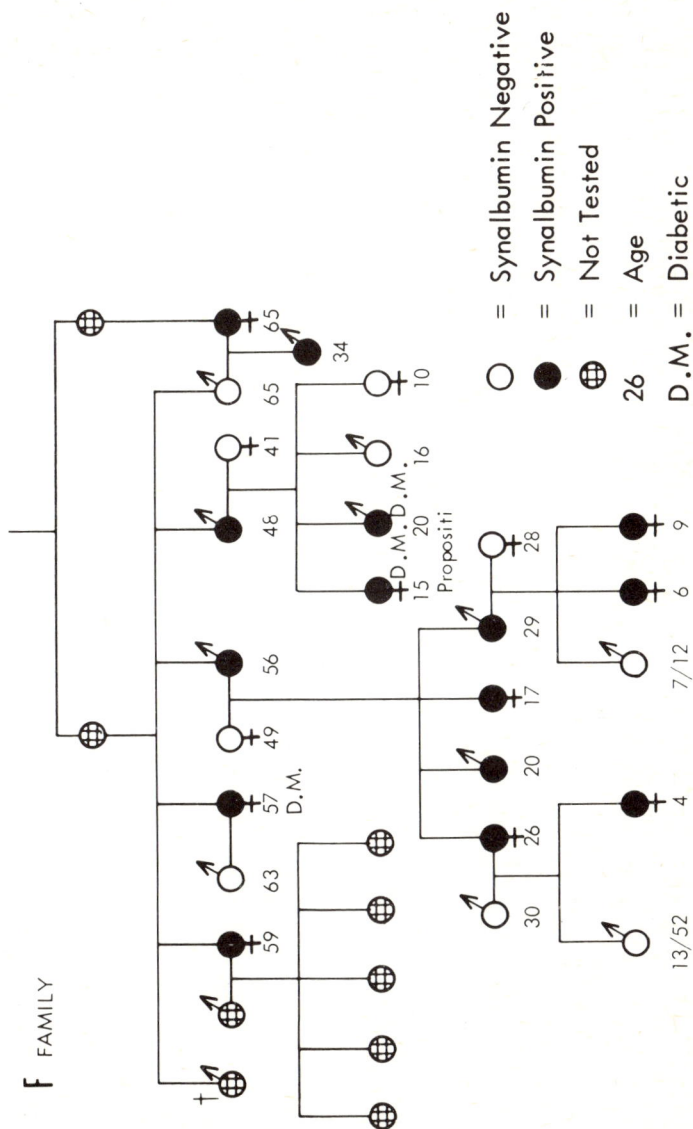

Figure 6.1 One large family pedigree showing the prevalence and inheritance of excessive synalbumin antagonsim through three generations.

seen that synalbumin positive mothers have cord bloods which are negative as well as positive and it is noteworthy that cord blood from two synalbumin negative women whose husbands were synalbumin positive were found to be synalbumin positive[74]. Therefore it seems that a child at birth can either be synalbumin positive (constituted as a diabetic) or synalbumin negative (normal).

Table 6.3 Studies on synalbumin antagonism of samples of cord blood associated with synalbumin positive and synalbumin negative mothers

	Cord blood	
Mothers	*Positive*	*Negative*
15+	9	6
12−	3*	9

Synalbumin positive +, antagonistic to insulin at 1.25%
Synalbumin negative −, non-antagonistic to insulin at 1.25%

*Two fathers +, one not tested

From the above investigations it appears that two phenotypes exist in the population: synalbumin positive and synalbumin negative, suggesting the inheritance of two alternative alleles at a single locus and that the division is not artificial. It also suggested that the homozygous and heterozygous members of the synalbumin positive group cannot be distinguished by the present rat diaphragm test. Alternatively, all the synalbumin positive individuals are heterozygotes, the homozygous state being lethal. Some support for this latter notion, which is the more attractive, derives from the high abortion rate in diabetic pregnancies[75] and the high perinatal mortality which is around 18%[76][77]. Thus when two diabetics marry only 67% of their offspring should be affected (synalbumin positive), the remaining 33% being normal (synalbumin negative). It is therefore of interest that Navarrete and Torres[78] found that in 55 subjects with two diabetic parents who previously had normal glucose tolerance, 67.2% showed abnormalities when tested by the triamcinolone glucose tolerance test (54.5% abnormal and 12.7% suspicious).

Whether or not a synalbumin positive individual constituted as an essential diabetic develops carbohydrate intolerance will depend on the degree of antagonism which can be further increased by environmental and physiological events *and* on the ability of the pancreas to withstand the challenge. This resilience of the pancreas will vary from family to family, like height for instance. Therefore in this sense the development of carbohydrate intolerance is multifactorial. Moreover it can be anticipated that the offspring of insulin-requiring diabetics who developed their carbohydrate intolerance in youth would give birth to children who, if they were going to develop overt diabetes, would do so much earlier than offspring of an obese diabetic whose pancreas withstood the challenge well into middle or even late life. In this way the observations of Simpson[79] can be explained: she found that there was a marked excess of overt diabetics among first degree relatives of diabetic patients whose age of onset was less than 20 years, compared with the relatives of patients with late age of onset of the condition.

In another report, 26 of 48 siblings and 5 of 9 parents of diabetics manifested increased synalbumin antagonism[35]; and in less extensive studies, increased antagonism has also been demonstrated in the following proportions of close relatives: 3 out of 7[80]; 4 out of 8[81]; 2 out of 4[41].

It is of particular interest that we have now been able to follow up the 63 members of the original six families studied for synalbumin antagonism between 1960 and 1962: 24 of these were synalbumin negative and none of these, twelve years later, have become diabetic; however, of the 39 synalbumin positive members there were only eight overtly diabetic up to 1962, and now 14 are clearly diabetic. The difference between 0 out of 24 v. 6 out of 31 is highly significant ($p = 0.03$). As far as we know, this is the first time that a prediction that a certain group of individuals may become diabetic has been made and subsequently upheld.

CLINICAL STUDIES

Many vascular complications of diabetes are no different from those seen in ageing people and they are dependent on the development of atheroma, for example angina of effort, or intermittent claudication, although they tend to occur at an earlier age and with more regularity in diabetic patients. Moreover, often in constituted diabetics such conditions are not infrequently found during the period before carbohydrate intolerance supervenes. We have already shown that 19 out of 28 unselected patients with cardiac infarction, who were not diabetic and who had no family history of the condition, had increased synalbumin antagonism to insulin when compared with six out of the same number of controls matched for sex and age[82]. Since then 40 patients with ischaemic heart disease have been studied, 19 males and 21 females. They were selected by a cardiologist on the basis of their strong family history of ischaemic heart disease or their unusual youth, many being well under 40 years of age. In this select picked group, 35 were synalbumin positive. Increased synalbumin antagonism was also noted in 10 out of 13 patients with angina pectoris and electrocardiographic changes of 'diffuse myocardial lesions' but with no evidence of diabetes mellitus[42]. In the same study 26 of 33 patients who had suffered a myocardial infarction in the past year also manifested increased synalbumin antagonism.

The relationship between the diabetic state in the mother and a tendency to deliver large babies has been known for many years and now it is well recognised that this tendency may be present even before diabetes mellitus is clinically manifest. A birthweight of more than 9.5 lb is now accepted as indicating prediabetes or potential diabetes in the mother. Of 23 women with a history of delivering an infant of this weight or more, 15 showed increased synalbumin antagonism (65%), as against 26 (36%) of 72 age-matched control subjects[83].

It is well-known that major congenital abnormalities are more common in the offspring of women suffering from diabetes[84]. Increased synalbumin antagonism was found in 24 of 36 mothers of children born with cleft palate and/or lip[85] and in 13 of 14 mothers delivered of babies with major spinal deformities and also in 15 of 18 mothers delivered of babies with major deformities of the extremities[86]. In the control group used for these studies only 14 of 50 women had increased antagonism.

Increased synalbumin antagonism has also been found in women with unexplained intra-uterine deaths during the third trimester of pregnancy[87]. However, increased synalbumin antagonism was also demonstrated in 5 out of 10 normal women during the third trimester of pregnancy[31] although these women did not deliver large babies nor had a family history of diabetes. Having regard for the fact that the antagonism is mediated by the pitiutary/adrenal system it is of course not surprising that the antagonism increases even in normal people during the third trimester of pregnancy when pitiutary/adrenal activity is so much increased.

These studies of synalbumin insulin antagonism are always carried out on muscle tissue. As already indicated, there is now considerable evidence that this antagonism does not affect insulin action on adipose tissue[37 41 88] — the conversion of carbohydrate to fat appears to be unimpeded. Therefore it has been postulated that the obese maturity-onset diabetics become obese because they are diabetic[27]. After the age of forty, newly diagnosed diabetics are on average appreciably overweight, the mean excess being almost 12% in each sex[89] although some weight loss has usually occurred before advice is sought. When ten grossly obese women ranging in age from 21 to 70 who were at least 25% overweight, were tested, excessive synalbumin insulin antagonism corresponding to that found in essential diabetics was found in six of them, although in one was there carbohydrate intolerance or a diabetic family history[90].

A number of studies have been carried out in patients with necrobiosis lipoidica diabeticorum who had no carbohydrate intolerance and in patients with granuloma annulare. All the former were found to be synalbumin positive, whereas 10 out of 14 were synalbumin positive in the latter group[91]. A close clinical association has been found between vitiligo and diabetes mellitus. Dawber[92] found that 4.8% of diabetics have vitiligo whilst others have demonstrated a high incidence of diabetes mellitus in the families of patients with vitiligo[93]. Of 21 patients with vitiligo 13 were synalbumin positive although they had no clinical evidence of carbohydrate intolerance[94].

It appears therefore that increased synalbumin antagonism can be found in a high percentage of conditions associated with diabetes mellitus and that it can be used as a marker of the prediabetic or latent diabetic stage of the condition.

BIOCHEMISTRY AND NATURE OF SYNALBUMIN

Early observations suggested that the synalbumin antagonist was of low molecular weight, was unlikely to be a free lipid, fatty acid or steroid type compound and was possibly a polypeptide[61]. Subsequent studies with Sephadex G-25 suggested that the antagonist had a molecular weight of less than 4000. There is an enzyme mainly concentrated in the liver which has a major specificity for the degradation of the insulin molecule. This purified enzyme called glutathione insulin transhydrogenase (GIT) catalyses the cleavage of the disulphide bonds of insulin by glutathione, resulting in the formation of the A- and B-chains of the insulin molecule[95]. The B-chain is a stable polypeptide of 3800 molecular weight with two SH groups per molecule, which in this reduced state is virtually insoluble in physiological buffers. Radiolabelled insulin in the presence of the purified enzyme GIT is cleaved into two radioactive components. These compounds, distinguished from the undegraded insulin at the origin by their electrophoretic migration in the α-2-globulin and albumin regions, were isolated and identified as the radio-active A- and B-chains respectively. It has also been established that B–chain migrating with the albumin electrophoretically is in fact bound to albumin and that this interaction permits solubilisation of the B-chain at physiological pH, whereas the A-chain is not associated with any serum macromolecule[96].

As there is extensive evidence indicating that inhibitors of a biochemical reaction are frequently structurally similar to the compounds which stimulate the reaction, it seemed likely that the B-chain might be capable of inhibiting the action of its parent molecular insulin; and since this chain is associated with albumin that it might be equated with the synalbumin insulin antagonist of human plasma. It has now been shown that when an albumin B-chain complex resulting from incubation of small amounts of B-chain with non-antagonistic albumin is assayed in the rat diaphragm system with insulin, marked antagonism to the hormone occurs[30]. There is now clear-cut evidence that the reduced B-chain of insulin combines with and rapidly disrupts the insulin molecule[97]. Moreover, reduced insulin B-chain complexed with albumin significantly increased blood glucose levels when injected into fasted normal rats and produced marked hyperglycaemia in rats fed on a high fat/high protein diet[98]. As with the synalbumin antagonist this last effect was dependent upon the integrity of the pituitary/adrenal system[99]. Studies which are continuing indicate that the synalbumin antagonist and the B-chain of insulin have a large number of physico-chemical similarities, notably molecular weight, the type of binding to albumin and ionic charge as well as biological and immunological activity (see page 171).

THE LIVER AND SYNALBUMIN PRODUCTION

Insulin secreted by the pancreas enters the portal vein and passes directly to the liver. In view of the presence of glutathione insulin transhydrogenase mentioned above, much of this insulin is broken down but as well as this it has been shown[100] that in a single passage through the liver 52% of injected labelled insulin is bound to the liver tissue. Thus it follows that the amount of insulin released into the circulation is dependent upon the liver; the half-time for the disappearance of labelled insulin from plasma in man was initially shown to be of the order of 40 min[101] but has since been found by a number of authors to be very much shorter and is now believed to be in the region of a few minutes only. That the peripheral insulin concentration is modified by the liver was well shown by Kaneto *et al.* in 1967[102]; a few minutes after stimulating the thoracic dorsal vagus nerve in a dog the insulin concentration in the portal vein plasma rose from 200 to 300 μu/ml, whereas no significant rise in peripheral vein insulin concentration could be shown; the change was only from 14μu/ml to 17 μu/ml.

In view of the above observations and that the liver is also the site of albumin formation, it seemed likely that the synalbumin was formed in this organ. Accordingly, experiments were undertaken to ascertain whether albumin from normal subjects initially devoid of antagonism through storage[51] could be rendered antagonistic again by perfusion with insulin through an isolated rat liver, using Miller's technique[103] as modified by Hems *et al*.[94] Albumin extracted after perfusion and tested at 1.25% on the rat hemidiaphragm showed significant antagonism to the insulin-stimulated glucose uptake. Albumin perfused through the liver without insulin, or albumin perfused with insulin through the glass apparatus without the liver, showed no insulin antagonism when similarly tested. The results are graphically represented in **Figures 6.2** and **6.3**. These observations[105] are compatible with the already considerable evidence that synalbumin and the B-chain of the insulin molecule are identical and strongly suggest that synalbumin (B-chain albumin) is produced in the liver. It is of interest in view of the work by Holcomb and Dulin[47] (mentioned on page 174) that all these re-extracted albumin preparations were non-antagonistic except the sample prepared from the perfusion medium containing insulin.

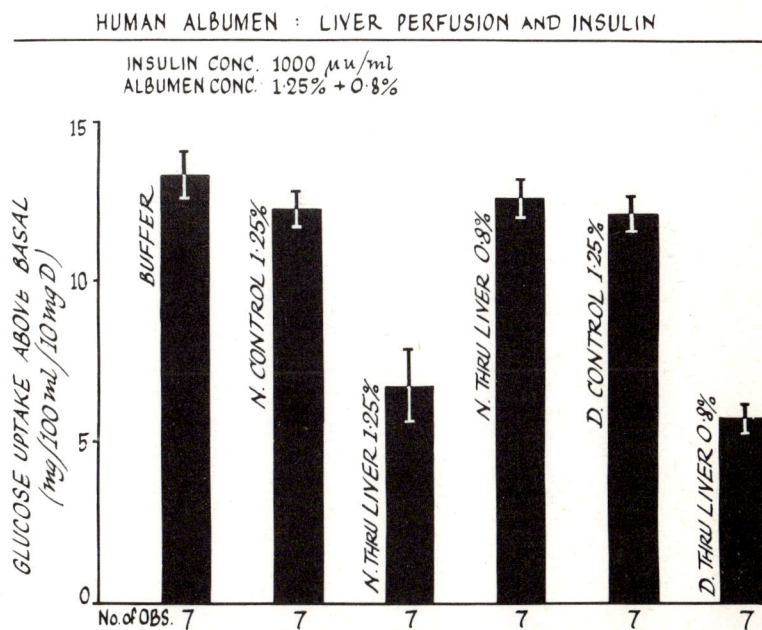

HUMAN ALBUMEN : LIVER PERFUSION AND INSULIN

Figure 6.2 Effect of insulin on glucose uptake of the rat diaphragm in the presence of albumin samples, tested at 1.25% and at 0.8% following liver perfusion with insulin.

HUMAN ALBUMEN AND LIVER PERFUSION minus INSULIN

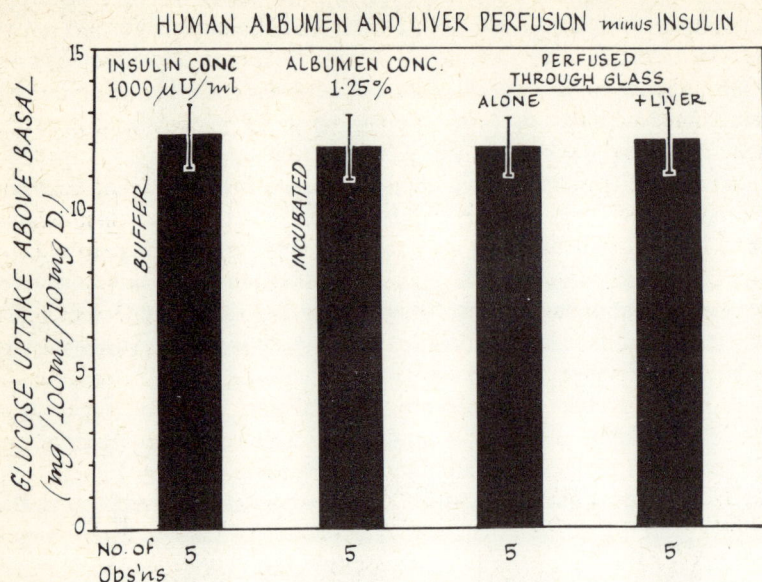

Figure 6.3 Similar states to Figure 6.2/but without insulin in the perfusing fluid of the isolated liver.

Although it would appear from these studies that synalbumin is produced from insulin in the liver and bound to the albumin there, the cause of the increased synalbumin antagonism of constituted diabetics is still quite unknown. Clearly several possibilities exist, but one of these is of particular interest, namely a subtle change in the characteristics of diabetic albumin, allowing it to carry larger amounts of B-chain, as was postulated in 1967[106].

As a result of the above studies, we measured the level of insulin antagonism achieved by albumin (stored for at least four weeks) from healthy normal subjects and diabetic patients after perfusion through the isolated rat liver. Albumin obtained from healthy controls, originally synalbumin negative, remains non-antagonistic to insulin at 1.25% after controlled incubation; it also does not become antagonistic after liver perfusion when tested at 0.8%. Similarly, albumin from diabetic patients (previously synalbumin positive) which has also lost its antagonism on storage continues to remain non-antagonistic at 1.25% after controlled incubation; however after liver perfusion it now becomes highly antagonistic when tested at 0.8%[107]. (see figure 6.2).

These findings indeed do suggest that diabetic and normal albumin differ in their ability to bind B-chain. If this difference in binding capacity can be confirmed, not only with radioactively labelled B-chain but also perhaps with other compounds, then this protein abnormality could be an important inherited characteristic and could be caused by an alteration in the base sequence of deoxyribonucleic acid (DNA).

A number of workers have attempted to confirm the original B-chain hypothesis of the synalbumin antagonist and it has been shown that both chemically and enzymatically derived reduced B-chain can render non-antagonistic albumin antagonistic again when measured not only by the *in vitro* bioassay[30], but also *in vivo*[99 108 109]. Immune rabbit gamma globulin prepared by injection of the heat-denatured supernatant fraction from antagonistic albumin preparations has been shown to inhibit the *in vitro* insulin antagonism of normal, diabetic and B-chain albumin[110]. Although so far B-chain has not been identified by chemical means in albumin preparations, similar ion exchange chromatographic behaviour[30], molecular weight[111 112], electrophoretic mobility[38 96] as well as its antagonistic activity, all lend support to the B-chain hypothesis. Until recently A- and B-chain immunoassay studies have been unsatisfactory and have not shown B-chain elevation in diabetic subjects[113 114].

A much more sensitive and precise radioimmunoassay for the A- and B-chains of insulin has now been perfected[115]. These assays are capable of detecting 200 pg/ml of A-chain and 100 pg/ml of B-chain using antibodies raised in guinea-pigs against the oxidised pork A-chain and against sulphonated pork B-chain. Now a number of studies have been performed using this technique. When insulin is perfused through the isolated liver in the presence of commercially produced human or bovine albumin, there is a fall in the insulin concentration over 30 min, with a corresponding rise in the A-chain and B-chain concentration. Moreover when either the A-chain or the B-chain is perfused through the liver separately or together without insulin, there is no change in the concentration of these chains[116]. These studies indicate that insulin is degraded into its A- and B-chains by glutathione and glutathione insulin transhydrogenase in the liver and that the chains so formed are not subject to further breakdown by proteolytic enzymes during the 30 min of the infusion period. These findings are again compatible with B-chain albumin formation in the liver.

Although as indicated on page 183 there are a very large number of chemical and physical similarities between the B-chain the synalbumin insulin antagonist, until the production of the immunoassay for B-chain it has not been possible to test this contention directly. This has now been done, for when albumin prepared from diabetic patients is oxidised with formic acid and hydrogen peroxide[117] and then chromatographed on

Dowex 50, it has been found that the material eluted under alkaline conditions and then recovered by lyophilisation reacts with the B-chain antibody in the immunoassay system[118]. When the same experiment is carried out using albumin prepared from normal volunteers the recovered material does not react with the same B-chain antibody, probably because the concentration is not sufficiently high. These observations are in line with our findings using the rat diaphragm assay that there is more antagonism associated with diabetic albumin than that from normal subjects.

DISCUSSION AND IMPLICATIONS

Normal subjects

Although many questions remain as yet unanswered regarding the role of synalbumin insulin antagonism in normal people, it is clear that it is closely associated with insulin metabolism and through this with that of carbohydrate, fat and protein. It is well-known that high concentrations of glucose in blood circulating through the pancreas stimulate insulin secretion by islets and that glucagon-like material from the gut enhances this effect. However, the effectiveness of any insulin secreted by the pancreas will depend on the activity of the liver. Undoubtedly this liver effect on insulin can mask actual increases in pancreatic secretion as illustrated by the work of Kaneto *et al*[102]: within a few minutes of stimulating the thoracic dorsal vagus nerve in a dog they found that the insulin concentration in portal venous plasma rose from 200 to 300 $\mu u/ml$, whereas there was no significant rise in the femoral vein insulin concentration (14–17 $\mu u/ml$). In other words, peripheral insulin concentration is monitored by the liver through the enzyme system GIT, which in turn has been shown to be controlled by adrenal cortical steroids and the pituitary gland[119], the effective agent being growth hormone[120]. It is therefore of interest that hypoglycaemia increases growth hormone output within a few minutes and that hyperglycaemia equally rapidly suppresses the secretion[121]. Thus there is evidence that part of the regulation blood-sugar may occur through the activity of this liver enzyme GIT, producing B-chain albumin (synalbumin) under the influence of growth hormone, namely mild hypoglycaemia→ growth hormone↑ + → GIT activity↑+ → more insulin degradation with less insulin coming out of liver and more B-chain formation which, attached to albumin, increases humoral insulin antagonism in the peripheral muscles, leading to a rise in blood sugar; the converse occurs with hyperglycaemia[29].

It is therefore of interest that it has already been shown that synalbumin antagonism changes during oral and intravenous glucose tolerance tests, the antagonism being inversely proportional to the glucose levels in the blood[64]. In these studies the blood sugar rose initially: accordingly we have recently studied intravenous tolbutamide tolerance tests where the immediate response is a fall in glucose levels[65]. In all the volunteers there was significantly increased antagonism to insulin at 30 min after the injection of tolbutamide. This corresponded to a peak hypoglycaemic response in these subjects. This increased antagonism at 30 min returned to pre-tolbutamide level at 90 min. These observations again showed a change in synalbumin related inversely to the blood sugar.

When considering growth hormone and its interrelation with the breakdown of insulin and insulin antagonism, it is worth remembering that a number of factors and situations are associated with increased growth hormone secretion[122]. These are fasting or starvation, sleep and exercise, and these situations would be associated with a decreased need for insulin action in the periphery. Again, other factors are situations associated with fever or stress, whether metabolic, physical or psychological and during anaesthesia and operations; these are well-known to be often associated with temporary or permanent carbohydrate intolerance.

It will be of interest to study the interrelationship of the gut hormones believed to be involved in insulin release from the pancreas and the enzyme system in the liver responsible for insulin breakdown. It seems reasonable to believe that stimulants of insulin release might, under ordinary circumstances, be associated with inhibition of the insulin breakdown mechanism, particularly if insulin is needed in the peripheral blood. Nevertheless, at present the short-term variations in the level of synalbumin appear to be due to changes in the pituitary/adrenal activity, which in turn is governed by acute changes in blood sugar as outlined above. Thus, regarding the control of the blood sugar level in normal subjects, the coarse adjustment is probably the output of insulin from the pancreas, whereas the fine adjustment occurs in the liver, mediated by its GIT system with breakdown of insulin and formation of B-chain resulting in variable antagonism to insulin in the peripheral muscles.

Diabetic patients: an aetiological concept of essential diabetes mellitus

The observations outlined above indicate that a fundamental abnormality in essential or idiopathic diabetes mellitus is increased synalbumin (B-chain albumin) antagonism to insulin–an exaggeration of the normal–which is present from birth and is apparently inherited by essential diabetics as an autosomal dominant character. The basic abnormality which results in this excessive circulating B-chain bound to albumin is not yet known but one possibility is that diabetic albumin and normal albumin differ in their ability

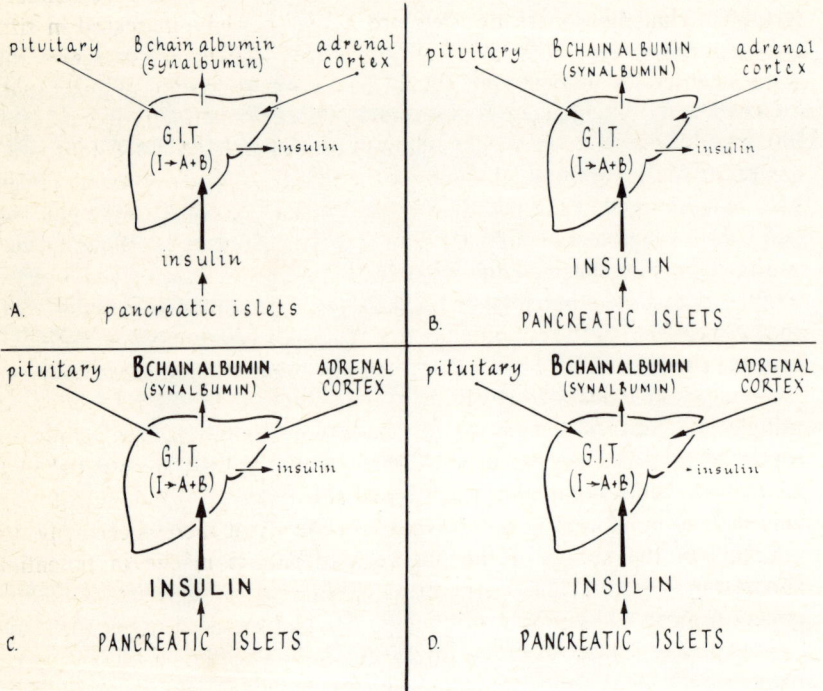

Figure 6.4 Diagrammatic representation of the formation and interaction between the synal-bumin insulin antagonist and insulin in relation to the aetiology of essential diabetes mellitus. (A) Normal.(B) Essential diabetes without carbohydrate intolerance.(C) Essential diabetes during pregnancy, infection and mental stress or during steroid administration, when no carbohydrate intolerance occurs. (D) Essential diabetes during pregnancy, infection and mental stress or during steroid administration, when carbohydrate intolerance occurs temporarily (latent diabetes). When the output of insulin from the pancreas falls further then essential diabetes becomes established, either with mild carbohydrate intolerance without ketosis ('obese' diabetics): or with severe carbohydrate intolerance associated with ketosis ('insulin-requiring' diabetics).

to bind B-chain, as indicated on page 187. In any event, it seemed clear that the synalbumin is made in the liver[107]. Thus in order to achieve increased B-chain albumin antagonism with even normal plasma insulin, more of this hormone must pass to the liver from the pancreas in potential or prediabetics; if there is also increased circulating insulin in this group then the input to the liver will be even greater. There is now reasonable agreement that the pancreatic islets are hypertrophied—increased in size and number—in prediabetics[11] [12], although there is still some disagreement as to whether immunoreactive insulin in plasma is normal or increased.

Cerasi and Luft injected glucose intravenously: a priming injection of 500 mg glucose/kg body weight followed immediately by a constant infusion of 20 mg/kg body weight/min for the next 60 min. They found, under these conditions, that the initial *peripheral* insulin response to the glucose load was less in potential diabetics than in normal subjects[5]. Although these results have now been widely quoted as indicating diminished insulin secretory capacity in potential or prediabetics, it is only an assumption that what is measured in the peripheral blood as insulin represents actual output from the pancreas. These studies have also been severely criticised because of the unphysiological nature of the insulin stimulus. Now, in relation to an oral glucose stimulus, several workers have reported finding high peripheral insulin levels and a greater insulin output *from* the liver in potential and chemical diabetics compared with normal subjects [123-127]. Undoubtedly the consensus of opinion at present favours the view that there is certainly no reduction in the ability of the pancreas to release insulin in potential diabetes and even in those with chemical diabetes, provided the stimulus, namely glucose, is given in a physiological way by mouth rather than un-physiologically by intravenous injection [125-127]. The above observations, made on peripheral blood, are in line with an increased input of insulin from the hypertrophied islets in potential diabetics, which is further supported by the reduction in basal splanchnic glucose production seen in prediabetes [128] [129].

The synalbumin antagonist is apparently dependent on the pituitary/adrenal system. It is reasonable therefore to suppose that the already increased antagonism will be even further increased under certain physiological and environmental conditions, notably the growth spurt, infection, pregnancy, the menopause, mental stress for any reason, or when adrenal cortical steroids are administered. These are times and situations which are well-known to precipitate carbohydrate intolerance in some susceptible individuals, or to aggravate this condition if it already exists. In this connection, as approximately 25% of the population appear to be constituted as diabetics, it is of interest that only about 25% of patients suffering from acute acromegaly, or Cushing's syndrome, actually develop overt carbohydrate intolerance. It is suggested that only those individuals constituted as diabetics exhibit carbohydrate intolerance when they develop the above conditions.

Moreover, it is possible to reproduce this kind of situation artificially by applying the steroid loaded glucose tolerance test. Rull and his colleagues [130] using such a stimulus measured the plasma insulin levels in first degree non-diabetic relatives of confirmed diabetics. Those subjects who gave a positive result indicating subclinical diabetes had plasma insulin levels which were lower than those in the relatives who gave a negative result. This observation was taken to indicate that the positive individuals were exhibiting a defect in the secretion of insulin. However in view of the increasing evidence that the breakdown of insulin by the enzyme GIT is under the control of the pituitary/adrenal system[120], a more logical explanation would be that in these circumstances more insulin is being broken down in the liver.

Whatever the environmental reason, in a certain, probably small, number of constituted diabetics, the β-cells of the pancreas which were initially hypertrophied and overactive ultimately begin to fail. The insulin production falls short of the requirements to a greater or lesser degree, leading to the well-recognised groups of insulin-requiring and obese diabetic patients respectively. However, it is of interest that Smith and Hall [131] studied carbohydrate intolerance in the very aged and found that out of 53 elderly non-diabetic volunteers aged 85–96, 19 had abnormalities of carbohydrate metabolism, and out of these, six had a clear explanation for this abnormality, such as 'mild organic brain syndrome', adrenal adenomata or raised blood urea. Thus in this extremely old group carbohydrate intolerance was found to occur in 13 out of 53, or 24.5%. It is therefore tempting to suggest that if life is prolonged sufficiently, all those who are constituted as diabetics may show abnormalities of carbohydrate intolerance. It is also of interest that in this group described above [131], in spite of a good insulin response to a glycaemic stimulus, these very old subjects have higher than normal blood sugar levels which suggests that the insulin secreted is either less biologically active or that this action is ineffective on account of antagonism which at least in this series was unrelated to obesity. Nevertheless, the phase before carbohydrate intolerance occurs is certainly very interesting and probably more important, especially in relation to obesity and the development of atherosclerosis.

Obesity and the diabetic constitution

The close association between obesity and diabetes is well-known and it has generally been assumed that obesity predisposes to the diabetic state. However, Medley [132] has examined obese people for susceptibility to diabetes by comparing intravenous glucose tolerance tests performed before and after an oral dose of prednisolone. His results strongly suggest that obesity *per se* is not a cause of diabetes, but whether it acts as a precipitating factor for those already predisposed to diabetes, or whether it is a consequence of the prediabetic state, is left open. As well as increased synalbumin (B-chain albumin) antagonism, several studies indicate that early juvenile and obese diabetics, including those with slight or no changes in carbohydrate intolerance, have certainly normal and sometimes increased amounts of available insulin or insulin activity [133-136]. There is also evidence that the synalbumin insulin antagonist which was detected using muscle does not affect insulin action on adipose tissue [37]; the conversion of carbohydrate to fat is not impeded.

We have therefore suggested that obese diabetics become obese because they are diabetic [27] – the action of insulin being deviated from muscle to fat. This notion is strengthened by the findings of Alp and Recant [41] who also showed that albumin had a stimulating action on fat, far greater than that achieved by any insulin carried with it. Moreover, synalbumin can have a lipogenic effect [137] and can significantly reduce the evoked potentials from the lateral feeding centre of the hypothalamus and prevent the immediate action of the injection of insulin [56]. This latter effect may account for the increased food intake (polyphagia) seen in diabetics. If it can be confirmed that many more people are constituted as diabetics than ever get carbohydrate intolerance, then it seems likely that one not uncommon cause of obesity in the normal population is the diabetic constitution [90]. It has already been shown that Pima Indians who developed marked carbohydrate intolerance were, up to four years previously, 25% heavier than their fellow tribesmen who did not become diabetic [138].

Atherosclerosis and diabetes

As already indicated (page 125), a large number of patients suffering from some of the effects of atheroma or atherosclerosis also have increased synalbumin antagonism. Moreover, it has been shown that many patients with cardiac infarction have increased amounts of circulating insulin [139] [140]. If the intima of the aorta responds to insulin in the same way as adipose tissue, then the effect of insulin in the blood will cause increased lipogenesis in arterial tissue [141] and may also stimulate cholesterol synthesis [142], as well as retarding lipolysis [143], the overall effect being to accelerate atherosclerosis in the larger vessels. The evidence incriminating insulin and carbohydrate in atherogenesis is strong and is discussed in detail by Stout and Vallance-

Owen [144]. Thus as we advance our knowledge of synalbumin and essential diabetes we may also hope to increase our understanding of ischaemic vascular disease.

There has always been doubt that exhaustion atrophy can occur in endocrine organs. However, structural changes have been found in the pancreatic islets of genetically obese rats from overwork [145]. At five weeks of age, the islets from these genetically obese rats were moderately hypertrophied; after that age the hypertrophied islets became more prominent, until 24 weeks of age, with accompanying degranulation of β-cells. The plasma insulin level also continued to increase during this period although the glucose level remained normal. At 52 weeks of age, both the plasma insulin and the triglyceride levels in the blood decreased and in the pancreas was observed proliferation of fibrous tissue and well granulated β-cells in the hypertrophied islets. In these rats at any rate the pancreatic islets which were in an active state during the development of obesity with hyperinsulinaemia and hyperlipidaemia eventually degenerated, became fibroblastic and produced less insulin[145]. Also, we have reported the case of a man who, following partial gastrectomy in 1950, developed severe hypoglycaemic attacks. For these attacks he had a distal pancreatectomy in 1966, when the islets were found to be increased in size with β-cell hyperplasia. Within six months however, the severe hypoglycaemic attacks returned; because of the strong family history and his own history, he was diagnosed as a constituted diabetic in February 1970 and this was confirmed at the same time by the finding of excess synalbumin antagonism to insulin. Two and a half years later in 1972, he developed overt diabetes and is presently being treated with the sulphonylurea drugs [146].

ON-GOING AND FUTURE STUDIES

As was indicated earlier, the basic abnormality in the diabetic constitution may be that diabetic albumin and normal albumin differ in their ability to bind B-chain. If this difference in binding capacity can be confirmed, then this protein abnormality could be an important inherited characteristic due to an alternation in the base sequence of deoxyribonucleic acid. Again, confirmation of this differential binding might allow a quicker and easier method of assessing increased synalbumin antagonism than the method of the isolated rat diaphragm which we presently employ; this would clearly allow study of many clinical problems at present considered to be associated with the diabetic state.

Preliminary studies[147] have not shown any qualitative difference in the amino acid composition of the acid and alkali hyrolysates of synalbumin positive and synalbumin negative samples on two-dimensional descending paper chromatography; also there have been no certain differences in the number of free thiol groups on such albumins when measured by polarographic methods [148]. Further studies have been carried out in which radiolabelled B-chain has been incubated with albumin and the complex subjected to isoelectric focusing [148]. The albumin complexes from diabetic and potential diabetic donors gave a different pattern from those of normal subjects [148]. Preliminary observations with polyacrylamide gel electrophoresis [149] suggest that albumin from diabetics and potential diabetics has a molecular weight greater by 4000 than normal albumin, which may indicate an extra B-chain molecule attached to the diabetic plasma protein [150]. If there are differences in the primary and/or secondary structure of albumin isolated from diabetics and non-diabetics, it is unlikely that this change can be corrected. However, it might be possible to modify the binding characteristics of albumin for B-chain. It is entirely possible that the extra B-chain might be displaced from its binding site on diabetic albumin by substances with a lower molecular weight, producing the situation found in normal subjects. In the first instance, it would be logical to try fragments from one or other end of the B-chain molecule itself. In short, if this differential binding capacity between diabetic and normal albumin can finally be confirmed, it opens up several possibilities: not only the better understanding of the constituted diabetic state with all this implies, but also a pointer towards a new form of treatment for the condition which could begin before the symptom of carbohydrate intolerance has developed.

References

1. Mering, J. von and Minkowski, O. (1889). Diabetes mellitus nach Pankreasextirpation. *Arch. Exp. Pathol. Pharmakol.,* **26,** 371
2. Banting, F. G. and Best, C. H. (1922). The internal secretion of the pancreas. *J. Lab. Clin. Med.,* **7,** 251
3. Seltzer, H. S. and Allen, E. W. (1963). Evidence that the primary lesion in diabetes mellitus is biochemical inertia of the beta cell. *J. Lab. Clin. Med.,* **62,** 1014
4. Cerasi, E. and Luft, R. (1963). Plasma insulin response to sustained hyperglycaemia induced by glucose infusion in normal subjects. *Lancet,* **ii,** 1359
5. Cerasi, E. and Luft, R. (1967). Further studies on healthy subjects with low and high insulin response to glucose infusion. *Acta Endocrinol. (Kbh),* **55,** 278
6. Goldner, M. G. and Clark, D. E. (1944). The insulin requirement of man after total pancreatectomy. *J. Clin. Endocrinol. Metab.,* **4,** 194
7. McCullagh, E. P., Cook, J. R. and Shirley, E. K. (1958). Diabetes following total pancreatectomy: clinical observation on 10 cases. *Diabetes,* **7,** 298
8. Whitfield, A. G. W., Crane, C. W., French, J. M. and Bayley, T. J. (1965). Life without a pancreas. *Lancet,* **i,** 675
9. MacLean, N. and Ogilvie, P. F. (1959). Observations on the pancreatic islet tissue of

young diabetic subjects. *Diabetes*, **8,** 83

10. Lazarus, S. S. and Volk, B. W. (1962). *The Pancreas in Human and Experimental Diabetes* (New York: Grune and Stratton)

11. Gepts, W. (1969). Does diabetes begin with insulin resistance or antagonism? Morphological aspects. In: *Diabetes*, 273 (I, Ostman, editor) (Amsterdam: Excerpta Medica).

12. Bloodworth, J. M. B. (1970). In: *Advances in Metabolic Disorders, Suppl. 1: Early Diabetes*, 260 (R. E. Camerini-Davalos and H. S. Cole, editors) (New York: Academic Press)

13. Vallance-Owen, J., Hurlock, B. and Please, N. W. (1955). Plasma-insulin activity in diabetes mellitus measured by the rat diaphragm technique. *Lancet*, **ii,** 583

14. Berson, S. A. and Yalow, R. S. (1962). Immunoassay of plasma insulin. In: Immunoassay of Hormones. *Ciba Fdn. Colloq. Endocrinol.*, **14,** 187

15. Yalow, R. S., Glock, S. M., Roth, J. and Berson, S. A. (1965). Plasma insulin and growth hormone levels in obesity and diabetes. *Ann. N. Y. Acad. Sci.*, **131,** 357

16. Danowski, T. S., Lombardo, Y. B., Mendelsohn, L. V., Corredor, D. G., Morgan, O. R. and Sabeh, G. (1969). Insulin patterns prior to and after onset of diabetes. *Metabolism*, **18,** 731

17. Fraser, R. (1970). Insulin in blood and urine. In: *Scientific Basis of Medicine*, 206 (London: Athlone Press)

18. Gemmill, G. L. (1941). The effect of glucose and of insulin on the metabolism of the isolated diaphragm of the rat. *Bull. Johns Hopkins Hosp.*, **68,** 329

19. Stadie, W. C. and Zapp, J. A. Jr (1947). The effect of insulin upon the synthesis of glycogen by rat diaphragm *in vitro*. *J. Biol. Chem.*, **179,** 55

20. Krahl, M. E. and Park, C. R. (1948). The uptake of glucose by the isolated diaphragm of normal and hypophysectomised rats. *J. Biol. Chem.*, **174,** 939

21. Vallance-Owen, J. (1956). Measurement of insulin activity in blood. *Diabetes*, **5,** 248

22. Vallance-Owen, J. and Wright, P. H. (1960). Assay of insulin in blood. *Physiol. Rev.*, **40,** 219

23. Vallance-Owen, J. and Hurlock, B. (1954). Estimation of plasma insulin by the rat diaphragm method. *Lancet*, **i,** 68

24. Gundersen, K. and Williams, R. H. (1960). Insulin antagonism in serum of untreated diabetes and in previously treated diabetes with ketoacidosis. *Proc. Soc. Exp. Biol. Med.*, **105,** 390

25. Vallance-Owen, J., Dennes, J. and Campbell, P. N. (1958). Insulin antagonism in plasma of diabetic patients and normal subjects. *Lancet*, **ii,** 336

26. Jackson, W. P. U. (1959). Prediabetes — a synthesis. *Postgrad. Med. J.*, **35,** 287

27. Vallance-Owen, J. and Lilley, M. D. (1961). Insulin antagonism in the plasma of obese diabetics and prediabetics. *Lancet*, **i,** 806

28. Vallance-Owen, J. (1962). Diabetes mellitus — causation. *Proc. Roy. Soc. Med.*, **55,** 207

29. Vallance-Owen, J. (1964). Synalbumin insulin antagonism and diabetes. *Ciba Fdn. Colloq. Endocrinol.*, **15,** 217

30. Ensinck, J. W., Mahler, R. J. and Vallance-Owen, J. (1965). Antagonism of insulin action on muscle by the albumin-bound B-chain of insulin. *Biochem. J.*, **94,** 150

31. Alp, H. and Recant, L. (1965). Studies of the insulin inhibitory effect of human plasma fractions. *J. Clin. Invest.*, **44,** 870

32. Davidson, M. B. and Goodner, C. J. (1966). Studies on the mechanism of insulin

antagonism by albumin in rat diaphragm. *Diabetes*, **15**, 373

33. Sherman, L. (1966). The Vallance-Owen ('synalbumin') insulin antagonist. Reproducibility and reliability of results in non-diabetic and diabetic humans. *Diabetes*, **15**, 149

34. Buse, M., McMaster, J. and Buse, J. (1966). Effect of human serum albumin on the glucose metabolism of isolated rat diaphragm and its response to insulin. *Diabetes*, **15**, 388

35. Ehrlich, R. M. and Martin, J. M. (1966). Presence of synalbumin antagonist in siblings of diabetic children. *Diabetes*, **15**, 400

36. Mahler, R. J. and Szabo, O. (1967). Effects of normal human albumin upon glucose uptake by the isolated rat diaphragm in the presence and absence of insulin. *Metabolism*, **16**, 853

37. Lowy, C., Blanshard, C. and Phear, D. (1961). Antagonism of insulin by albumin. *Lancet*, **i**, 802

38. Jervell, J. (1965). The antagonistic effect of human plasma albumin on the insulin-stimulated glucose uptake of the isolated rat diaphragm. *Acta Physiol.* **65**, 33

39. Holcomb, G. N. and Dulin, W. E. (1968). Studies on the nature of synalbumin insulin antagonist. *Proc. Soc. Exp. Biol. Med.*, **128**, 154

40. Alp, H., Chaplin, H. and Recant, L. (1966). Partial purification of an insulin inhibitor from human albumin fractions. *J. Clin. Endocrinol.*, **26**, 340

41. Alp, H. and Recant, L. (1964). Effect of the insulin-inhibitory albumin fraction from normal and diabetic subjects on adipose tissue. *Metabolism*, **13**, 609

42. Kammerer, L., Bretan, M., Nemesanszky, L., Jakali, L. and Virag, S. (1966). A plasma-synalbumin antagonist. *Lancet*, **ii**, 1366

43. Kimberling, W. J., Coneally, P. M. and Merritt, A. D. (1966). Insulin antagonism in families with juvenile diabetes. *3rd Int. Congr. Human Genetics, Chicago, 5–10 September*

44. Keen, H. (1963). Properties of human 'albumin'. A metabolic study of albumin extracts from normal and diabetic plasma. *Diabetes*, **12**, 406

45. Cameron, J. S., Keen, H. and Menzinger, J. (1964). Insulin activity of normal plasma and plasma acid-ethanol extracts. *Lancet*, **ii**, 74

46. Ensinck, J. W., Poffenbarger, P. L., Hogan, R. A. and Williams, R. H. (1967). Studies of insulin antagonism. I. An artefactual antagonist to insulin and plasma non-suppressible insulin-like activity occurring in preparation of 'albumin'. *Diabetes*, **16**, 289

47. Holcomb, G. N. and Dulin, W. E. (1973). The nature of the artefactual synalbumin insulin antagonist. *Diabetologia*, **9**, 509

48. Debro, J., Tarver, H. and Korner, A. (1957). The determination of serum albumin by a new method. *J. Lab. Clin. Med.*, **50**, 728

49. Lilley, M. D. and Vallance-Owen, J. (1961). A factor associated with 'Visking' tubing capable of antagonising insulin. *Nature (Lond.).* **190**, 1196

50. Fernandez, A., Sobel, C. and Goldenberg, H. (1966). An improved method for determination of serum albumin and globulin. *Clin. Chem.*, **12**, 194

51. Vallance-Owen, J. and McMaster, D. (1968). Serum-albumin and insulin antagonism. *Lancet*, **ii**, 1192

52. Bajaj, J. S., Bansal, D. D., Jindal, H. O. and Garg, S. K. (1974). A comparative evaluation of the methods for the study of synalbumin insulin antagonism. *Acta Diabetol. Latina* (in press)

53. Vallance-Owen, J. and McMaster, D. (1973). Conservation of the synalbumin insulin

antagonist. *Clin. Chim. Acta,* **45,** 99

54. McMaster, D., Dunn, P. G. M. and Vallance-Owen, J. (1974). Unpublished observations

55. Bajaj, J. S. and Vallance-Owen, J. (1974). Fenfluramine and glucose uptake by muscle. *Horm. Metab. Res.,* **6,** 85

56. Chhina, G. S. and Bajaj, J. S. (1972). Nervous regulation of glucose homeostasis. In: *Insulin and Metabolism,* 155 (J. S. Bajaj, editor) (Diabetic Association of India)

57. Bajaj, J. S., Bansal, D. D. and Jindal, H. O. (1974). Synalbumin insulin antagonism and protein synthesis. *Acta Diabetol. Latina* (in press).

58. Dunn, P. G. M., McMaster, D. and Vallance-Owen, J. (1974). Unpublished observations

59. Kammerer, L., Bretan, M. and Steingaszner, O. (1970). Effects of the synalbumin insulin antagonist *in vivo;* its examination with a new method. *Diabetologia,* **6,** 473

60. Vallance-Owen, J., Dennes, J. and Campbell, P. N. (1958). The nature of the synalbumin insulin antagonist associated with plasma albumin. *Lancet,* **ii,** 696

61. Vallance-Owen, J. and Lilley, M. D. (1961). An insulin antagonist associated with plasma albumin. *Lancet,* **i,** 804

62. Bornstein, J. and Park, C. R. (1953). Inhibition of glucose uptake by serum of diabetic rats. *J. Biol. Chem.,* **205,** 503

63. Vallance-Owen, J. and Lukens, F. D. W. (1957). Studies on insulin antagonism in plasma. **60,** 625

64. Jervell, J. and Vallance-Owen, J. (1967). Variations in synalbumin insulin antagonism during glucose tolerance tests. *Lancet,* **ii,** 1253

65. Bajaj, J. S. and Vallance-Owen, J. (1971). Variations in synalbumin insulin antagonism during tolbutamide tolerance test. *Diabetologia,* **7,** 133

66. Pincus, G. and White, P. (1933). On the inheritance of diabetes mellitus. I. An analysis of 675 family histories. *Amer. J. Med. Sci.,* **186,** 1

67. Steinberg, A. G. (1959). The genetics of diabetes: a review. *Ann. N. Y. Acad. Sci.,* **82,** 197

68. Steinberg, A. G. (1965). Genetics and diabetes. In: *On the Nature and Treatment of Diabetes,* 601 (B. S. Leibel and G. A. Wrenshall, editors) (Amsterdam: Excerpta Medica)

69. Motulsky, A. G. (1962). Genetics and endocrinology. In: *Textbook of Endocrinology,* 1107, 3rd edition (R. H. Williams, editor) (Philadelphia: Saunders)

70. Vallance-Owen, J. (1966). The inheritance of essential diabetes mellitus from studies of the synalbumin insulin antagonist. *Diabetologia,* **2,** 248

71. Bajaj, J. S., McMaster, D. and Vallance-Owen, J. (1971). Further studies on the incidence of the synalbumin insulin antagonist. *Diabetologia,* **7,** 136

72. Bajaj, J. S. (1970). Plasma-insulin in diabetes. *Lancet,* **ii,** 1395

73. Bajaj, J. S., Bansal, D. D. and Gupta, S. K. (1970). *Report on the enquiry 'Studies on the SIA as a genetic marker of diabetes mellitus in the Indian population'* (Indian Council of Medical Research)

74. Vallance-Owen, J. (1968). Synalbumin insulin antagonist. *Postgrad. Med. J.* (Festschrift for Sir John McMichael) **44,** 117

75. Joslin, E., Root, H., White, P. and Marble, A. (1958). *United States Department of Health, Education and Welfare, Public Health Service, Vital Statistics Special Report,* **39,** 705

76. Miller, M. (1965). Diabetic pregnancy and foetal survival in a large metropolitan area. In: *On the Nature and Treatment of Diabetes,* 714 (B. S. Leibel and G. A. Wrenshall,

editors) (Amsterdam: Excerpta Medica)

77. Osler, M. (1965). Structural and chemical changes in infants of diabetic and pre-diabetic mothers. In: *On the Nature and Treatment of Diabetes*, 962 (B. S. Leibel and G. A. Wrenshall, editors) (Amsterdam: Excerpta Medica)

78. Navarrete, V. N. and Torres, I. H. (1967). Triamcinolone provocative test in offspring of two diabetic parents. *Diabetes,* **16,** 57

79. Simpson, N. E. (1967). Multifactorial inheritance: a possible hypothesis for diabetes. *Diabetes,* **13,** 462

80. Levin, M. E.and Recant, L. (1967). Hormonal and biochemical studies in a diabetic family. *Ann. Int. Med.,* **66,** 69

81. Ashton, W. L. (1965). The glucose uptake induced in rat hemidiaphragms in the presence of human albumin. *J. Endocrinol.* **33,** 103

82. Vallance-Owen, J. and Ashton, W. L. (1963). Inheritance of essential diabetes mellitus from studies of the synalbumin insulin antagonist. *Diabetes,* **12,** 356

83. Vallance-Owen, J., McMaster, D. and Bajaj, J. S. (1973). Maternal synalbumin antagonism and large babies. *Lancet,* **ii,** 358

84. Pederson, L. M., Tygstrup, L. and Pederson, J. (1964). Congenital malformations in newborn infants of diabetic women. *Lancet,* **i,** 1124

85. Vallance-Owen, J., Braithwaite, F. and Wilson, J. S. P. (1967). Cleft lip and palate deformities and insulin antagonism. *Lancet,* **ii,** 912

86. Wilson, J. S. P. and Vallance-Owen, J. (1966). Congenital deformities and insulin antagonism. *Lancet,* **ii,** 940

87. Connon, J. J. (1969). Unexplained foetal deaths and synalbumin antagonism. *Diabetologia,* **5,** 188

88. Kipnis, D. M. and Stein, M. F. (1964). Insulin antagonism: fundamental considerations. *Ciba Fnd. Colloq. Endocrinol.,* **15,** 156

89. Pyke, D. A. (1959). Aetiological factors in diabetes. *Postgrad. Med. J.,* **35,** 261

90. Vallance-Owen, J. (1965). Synalbumin insulin antagonism in obesity and maturity-onset diabetes mellitus. *Ann. N. Y. Acad. Sci.,* **131,** 315

91. Vallance-Owen, J. (1969). Diabetes mellitus. *Brit. J. Dermatol.* **81,** 9

92. Dawber, R. P. R. (1968). Vitiligo in mature-onset diabetes. *Brit. J. Dermatol.,* **80,** 275

93. Cunliffe, W. J., Hall, R., Newell, D. J. and Stevenson, C. J. (1968). Vitiligo, thyroid disease and auto-immunity. *Brit. J. Dermatol.,* **80,** 135

94. Dawber, R. P. R., Bleehan, S. S. and Vallance-Owen, J. (1971). Vitiligo and diabetes mellitus. *Brit. J. Dermatol.,* **84,** 600

95. Tomizawa, H. H. and Halsey, Y. D. (1959). Isolation of an insulin-depending enzyme from beef liver. *J. Biol. Chem.,* **234,** 307

96. Ensinck, J. W., Coombs, C. J., Williams, R. H. and Vallance-Owen, J. (1964). Studies *in vitro* of the transport of the A and B chains of insulin in serum. *J. Biol. Chem.,* **239,** 3377

97. Zahn, H., Gutte, B. and Gattner, H. G. (1968). Reaction of reduced A- and B-chain with insulin. *Diabetologia,* **4,** 118

98. Alburn, H. E. and Fenichel, R. C. (167). Hyperglycaemia induced by insulin B-chain in dietary diabetes in rats. *Nature (Lond.).* **213,** 515

99. Fenichel, R. L., Bechmann, W. M. and Alburn, H. E. (1968). Pituitary and adrenal influence on reduced insulin B-chain induced hyperglycaemia. *Diabetes,* **17,** 67

100. Madison, L. L., Combes, B., Unger, R. H. and Kaplan, N. (1959). The relationship between the mechanism of action of the sulfonylureas and the secretion of insulin into

the portal circulation. *Ann. N. Y. Acad. Sci.*, **74**, 548

101. Berson, S. A. and Yalow, R. S. (1965). Some current controversies in diabetes research. The Banting Memorial Lecture, 1965. *Diabetes*, **14**, 549

102. Kaneto, A., Kosaka, K. and Nakao, K. (1967). Effects of stimulation of the vagus nerve on insulin secretion. *Endocrinology*, **80**, 530

103. Miller, L. L., Bly, C. G., Watson, M. L. and Bale, W. F. (1951). The dominant role of the liver in plasma protein synthesis. *J. Exp. Med.*, **94**, 431

104. Hems, R., Ross, B. D., Berry, M. N. and Krebs, H. A. (1966). Gluconeogenesis in the perfused rat liver. *Biochem. J.*, **101**, 284

105. Bajaj, J. S. and Vallance-Owen, J. (1974). Unpublished observations

106. Vallance-Owen, J. (1967). Current views on the aetiology of diabetes mellitus. In: *Modern Trends in Endocrinology*, 152, 3rd edition (H. Gardiner-Hill, editor) (London: Butterworths)

107. Bajaj, J. S. and Vallance-Owen, J. (1971). Insulin antagonism to normal and diabetic albumin after liver perfusion. *Lancet*, **i**, 16

108. Mahler, R. J., Szabo, O. and Penhos, J. C. (1968). Antagonism to insulin action on the perfused hind limb of the rat by a reduced insulin B-chain—albumin complex. *Diabetes*, **17**, 1

109. Fenichel, R. L., Bechmann, W. H. and Alburn, H. E. (1966). Inhibition of insulin activity in mitochondrial systems and in normal rats by reduced insulin B-chain—albumin complex. *Biochemistry*, **53**, 461

110. Vallance-Owen, J. (1968). Synalbumin insulin antagonism. *Folio Endocrinol. Jap.*, **44**, 455

111. Vallance-Owen, J. (1965). Insulin antagonists. In: *On the Nature and Treatment of Diabetes*, 340 (B. S. Leibel and G. A. Wrenshall, editors) (Amsterdam: Excerpta Medica)

112. Young, D. A. B. (1967). A serum inhibitor of insulin action on muscle. I. Its detection and properties. *Diabetologia*, **3**, 287

113. Meek, J. C., Doffing, K. M. and Bollinger, R. E. (1968). Radioimmunoassay of insulin A and B chains in normal and diabetic human plasma. *Diabetes*, **17**, 61

114. Varandani, P. T. (1968). Plasma concentrations of A and B chains of insulin in non-diabetic, diabetic and high risk potential diabetic subjects. *Diabetes*, **17**, 547

115. Bansal, D. D., Buchanan, K. D., Bajaj, J. S. and Vallance-Owen, J. (1973). Sensitive and precise radioimmunoassays for the A- and B-chains of insulin. *8th Congr. Int. Diabetes Fed., Brussels 1973, Int. Congr. Series 28014* (Amsterdam: Excerpta Medica)

116. Bansal, D. D. and Vallance-Owen, J. (1974). Unpublished observations

117. Sanger, F. (1949). Fractionation of oxidised insulin. *Biochem. J.*, **44**, 126

118. Vallance-Owen, J. and Bansal, D. D. (1974). Unpublished observations

119. Elgee, M. J. and Williams, R. H. (1955). Fate of insulin in altered metabolic states. *Diabetes*, **4**, 8

120. Costiner, E., Milcu, St M., Oprescu, M. and Simionescu, L. (1970). The anti-insulinic activity of the growth hormone. *Abstr. 7th Congr. Int. Diabetes Fed.* (Amsterdam: Excerpta Medica)

121. Roth, J., Glick, S. M., Yalow, R. S. and Berson S. A. (1963). Secretion of human growth hormone: physiologic and experimental modification. *Metabolism*, **12**, 577

122. Sonksen, P. H. (1974). Clinical applications of growth hormone assays. *J. Roy. Coll. Phys. Lond.*, **8**, 220

123. Paulsen, E. P., Richenderfer, L. and Ginsberg-Fellner, F. (1968). Plasma glucose, free fatty acids and immunoreactive insulin in sixty-six obese children. *Diabetes*, **17**, 261

124. Khurana, R. C., Robin, J. A., Jung, Y., Corredor, D. G., Gonzalez, A., Sunder, J. H. and Danowski, T. S. (1971). Hyperinsulinemia after tolbutamide in mild glucose intolerance. *Horm. Metab. Res.*, **3**, 233

125. Danowski, T. S., Lombardo, Y. B., Mendelsohn, L. V., Corredor, D. G., Morgan, C. R. and Sabeh, G. (1969). Insulin patterns prior to and after onset of diabetes. *Metabolism*, **18**, 731

126. Reaven, G. M., Shen, S. W., Silvers, A. and Farquhar, J. W. (1971). Is there a delay in the plasma insulin response of patients with chemical diabetes mellitus? *Diabetes*, **20**, 416

127. Jackson, W. P. U., van Miegham, W. and Keller, P. (1972). Insulin excess as the initial lesion in diabetes. *Lancet*, **1**, 1040

128. Cerasi, E., Wahren, J., Luft, R. and Felig, P. (1973). Hepatic sensitivity to endogenous insulin in prediabetes. *Diabetologia*, **9**, 63

129. Wahren, J., Felig, P., Cerasi, E., Luft, R. and Hendler, R. (1973). Splanchnic glucose production and its regulation in healthy monozygotic twins of diabetics. *Clin. Sci.*, **44**, 493

130. Rulle, J. A., Conn, J. W., Floyd, J. C. Jr and Fajans, S. S. (1970). Levels of plasma insulin during cortisone glucose tolerance tests in 'nondiabetic' relatives of diabetic patients. *Diabetes*, **19**, 1

131. Smith, M. J. and Hall, M. R. P. (1973). Carbohydrate tolerance in the very aged. *Diabetologia*, **9**, 387

132. Medley, D. R. K. (1965). The relationship between diabetes and obesity: the study of susceptibility to diabetes in obese people. *Quart. J. Med.*, **34**, 111

133. Yalow, R. S. and Berson, S. A. (1960). Plasma insulin concentration in non-diabetic and early diabetic subjects; determinations by a new sensitive immunoassay technique. *Diabetes*, **9**, 254

134. Yalow, R. S. and Berson, S. A. (1961). Immunoassay of plasma insulin in man. *Diabetes*, **10**, 339

135. Steinke, J., Soeldner, J. S., Camerini-Davalos, R. A. and Renold, A. E. (1963). Studies on serum insulin-like activity in pre-diabetes and early overt diabetes. *Diabetes*, **15**, 502

136. Hales, C. N., Walker, J. B., Garland, P. B. and Randle, P. J. (1965). Fasting plasma concentrations of insulin, non-esterified fatty acids, glycerol and glucose in the early detection of diabetes mellitus. *Lancet*, **i**, 65

137. Biener, J. and Vallance-Owen, J. (1969). Effect of synalbumin on lipolysis. *Lancet*, **ii**, 1390

138. Ingelfinger, J. A., Bennett, P. H., Kamenetzky, S. A., Savage, P. J., Dippe, S. E. and Miller, M. (1973). Characteristics of Pima Indians prior to the development of marked glucose intolerance. *Diabetes*, **22**, 289

139. Peters, N. and Hales, C. N. (1965). Plasma insulin concentration after myocardial infarction. Lancet, **i**, 1144

140. Nikkila, E. A., Tatu, A. M., Vesenne, M. and Pelkonen, R. (1965). Plasma insulin in coronary heart disease: response to oral and intravenous glucose and to tolbutamide. *Lancet*, **ii**, 508

141. Stout, R. W. (1968). Insulin stimulated lipogenesis in arterial tissue in relation to diabetes and atheroma. *Lancet*, **ii**, 702

142. Stout, R. W. (1969). Insulin stimulation of cholesterol synthesis by arterial tissue. *Lancet*, **ii**, 467

143. Mahler, R. J. (1965). Diabetes and arterial lipids. *Quart. J. Med.*, **34**, 484

144. Stout, R. W. and Vallance-Owen, J. (1969). Insulin and atheroma. *Lancet,* **i,** 1078
145. Shino, A., Matsuo, T., Iwatsuka, H. and Suzuoki, Z. (1973). Structural changes of pancreatic islets in genetically obese rats. *Diabetologia,* **9,** 412
146. Vallance-Owen, J. and Robb, J. D. A. (1974). Spontaneous hypoglycaemia and diabetes mellitus. *Irish J. Med. Sci.,* **143,** 21
147. Bajaj, J. S., Bansal, D. D., Garg, S. K. and Bansal, N. (1973). Amino acid analysis of synalbumin. *Acta Diabetol. Latina,* **10,** 1061
148. Asquith, R. S., Otterburn, M. S. and Francis, T. T. (1974). Unpublished observations
149. Dunker, A. K. and Rueckert, R. R. (1969). Observations on molecular weight determinations non polyarcrylamide gel. *J. Biol. Chem.,* **244,** 5074
150. Taylor, W. H. and Vallance-Owen, J. (1974). Unpublished observations

INDEX

Abdominal pain 104, 105
Acetylcholine 66
N-Acetylglucosamine 42, 44, 45
Achlorhydria 79
ACTH 47
Activation of proinsulin 4
Adenylcyclase 40–42, 48–50, 53, 56
Adipose tissue 150, 153, 158, 182
Adrenaline 41, 46, 48
Affinity chromatography 71
Alanine 49, 76, 94, 96, 97, 99, 100, 103, 105, 110, 114
Albumin
  fat stimulation 193
  insulin antagonism of plasma 173–176, 194, 195
  isolation 175
  synalbumin formation 185
Aldose reductase 145
Alloxan diabetes 137, 138, 146, 151
Amino acids
  hypoaminoacidaemia 100
  insulin release 48–50
  see also Amino acids by name
2–Aminobicyclo (2,2,1) heptane-2-carboxylic acid (BCH) 50
Animal models 138–140
  rat diaphragm model 172, 173, 176 180, 182, 185
Antagonism, plasma/insulin 172–175 178–180
  pregnancy 182
Antibodies
  pancreatic glucagon 71, 77
  secr,etin 73, 74
Aorta (atherosclerosis) 127, 128, 145, 146, 193
Arginine 49, 55, 76
Artefactual insual antagonist 174, 175
Artery (atherosclerosis) 125, 139–148
  cholesterol synthesis 141, 143
  insulin activity 146, 147
  phospholipid synthesis 142
Atherosclerosis 125, 127–148, 193 194
ATP concentration, islet 45

Autonomic neuropathy 79
Autophagy, granule 12

Barium meal (gastro-paresis) 79
Basal insulin secretion 137
Big, big insulin 16
Biosynthesis 1–14
  β-cell 6–8
  regulation 11–14, 50
Birthweight, offspring of diabetics 182
Brownian motion of granules 34

Caffeine 49, 50, 55
Calcium involvement 39, 40
Carbohydrate content of diet 136, 137, 155, 156, 172
Carbohydrate intolerance 174, 178, 181, 191–193

Carbohydrate metabolism 128
Carboxypeptidase B 8, 10
Cardiovascular disease 181, 182, 193
^{14}C-arginine 10, 11
Catecholamines 106, 150
β-Cells 1, 3, 6–8, 10, 12
  anatomy 31–37
  cyclic AMP concentrations 40–42
  glucoreceptor 42–46
  membrane electrical activity measurement 39
  secretory activity 21–23
Cell web 36
p-Chloromercuribenzoate (PCMB) 50
Cholecystokinin-pancreozymin (CCK–PZ) 67, 68, 70, 74, 82
Cholesterol 132, 133, 134, 136, 139, 142, 143, 144, 147, 156, 193
  hypecholesterolaemia 132, 136, 149, 155, 156
Chromatography, affinity 71
Chylomicrons 144, 148, 156, 157
Chylomicronanaemia 149, 156, 157
Chymotrypsin 8
Circulating C-peptide levels 21–24
Circulating insulin antibodies 21–23
Circulating proinsulin levels 16–18
Cirrhosis 117, 118

Cleavage of proinsulin 6, 8, 13, 14
Colchicine 35, 37
Congenital abnormalities, offspring of diabetics 182
Congenital total lipodystrophy 157, 158
Connecting peptide segment (CP) 32
Connective tissue protein 142
Cord blood preparations 180
Cori cycle 94
Coronary atherosclerosis 128
C-peptide 4, 5, 8, 11
  amino acid sequence 13
  circulation levels 21–24
  C-peptide reactivity (CPR) 20, 21–23
  metabolism 21
  serum 20
  structure and properties 13, 14
Crystallization of insulin 11
Cyclic 3',5'-AMP 40–42, 47
  dependent protein kinase 41, 42
  glucagon 70
  insulin activity, artery 146
  phosphodiesterase 40–42, 49, 53 55
  secretion of insulin 54–56
Cycloheximide 51
Cytochalasin B 35, 36

Deficiency
  insulin 77, 108, 150
  post-herapin lipolytic activity (PHLA) 151–154, 158
Depancreatization 171
Diabetic diarrhoea 79, 80
Diazoxide 41, 49, 50
Diet 136, 137, 155, 156
Diguanides 81
Duration of diabetes (atherosclerosis) 131

Electron microscopy 66
  β-cells 32, 33, 34, 35
  smooth muscle cells (lipoprotein) 142, 143
Endopeptidase 10
Entero-insular axis 64, 83
Esterification
  cholesterol 142, 143
  free fatty acid 150
Exercise, glucose homeostasis in 103–106, 112, 113

Exocrine pancreatic insufficiency 76, 80
Exocytosis 33–37, 43

Familial hypercholesterolaemia 156
Familial hypertriglyceridaemia 153, 154
Fasting
  hyperglycaemia 108–111, 149, 157 158
  hypoglycaemia 23
  post-absorptive 93–96
  proinsulin levels 16
Fatty acids 50, 112, 139
  cholesterol ester 142, 143
  synthesis 143
Fatty liver 150
Fatty streak lesions 140
Fenfluramine 176
Fernandez method of albumin isolation 175
Foam (smooth muscle) cells 140, 141
Framingham study 129, 132, 135
Free fatty acids 112, 113, 150, 151

Galactose 43, 44
Gastric inhibitory polypeptide (GIP) 48, 69, 72, 75, 78, 82
Gastrin 67, 68, 70, 74, 75, 77, 82
Gastrointestinal hormones 47, 48, 72–77
Gastro-paresis 79
Gel filtration 14–16
  glucagon-like immunoreactivity (GLI) 71
Genetic transmission of essential diabetes mellitus 178, 179
Glibenclamide 41, 80
Glucagon 47, 55, 56, 66, 69, 70, 74, 75, 76, 77, 82, 83, 102
Glucagon-like immunoreactivity (GLI) 70, 71, 74, 77, 78, 82
Glucagonomas 78
Gluconeogenesis 94–97, 100, 103, 105, 106, 110, 111, 112, 114
Glucoreceptors 42–46, 52, 53, 55
Glucosamine 45
Glucose
  balance 95, 97, 98
  C-peptide reactivity 23
  fasting 94–100
  glucoreceptor 42–46

hepatic production 94, 96, 97,
    109
homeostasis 96, 97, 99, 100–106,
    175
ingestion 97–100
insulin release 11, 12, 35, 38, 39,
    40 42–48, 55, 76, 77, 80, 102
intolerance 106–108, 153
obesity 113, 114
oral; *see* Oral glucose
polyol pathway 145, 146
proinsulin biosynthesis 57
secretin 74
tolerance 51, 56, 65, 106–108, 136,
    149
tolerance tests 107, 127, 128, 177
    180, 188, 192, 193
transport in gut 80, 81
uptake, effect of insulin on 185, 186
Glutathione insulin transhydrogenase
    (GIT) 183, 184, 186, 187, 192
Glyceraldehyde 12, 43, 46
Glycerol 114
Glycogen 94
    liver content 96
Glycogenolysis 94–96, 97, 99, 103,
    105, 106, 111
Glycolysis 55
Glycosuria 128
Golgi complex 6–8, 10, 11, 32
Granule autophagy 12
β-Granule formation 10, 11
Granule translocation system 34, 36,
    43
Granuloma annulare 183
Growth hormone 186, 189
Gut 63, 64, 79–81

Heparin (PHLA) 151–153
Hepatic release of glucose 94
Hepatitis, viral 116, 117
Hepatogenous diabetes 117, 118
Heterozygosity 180
High density lipoprotein 144
Homozygosity 178
Hormones 48, 49, 54
    fasting 93–95, 97
    gastrointestinal 66, 67, 72–77
    growth 188, 189
    synalbumin antagonism 176, 177
Hypercholesterolaemia 132, 136, 149,
    155, 156
Hyperglucagonaemia 118

Hyperglycaemia 56, 70, 73, 108
    atherosclerosis 130, 134, 135, 139,
        151–153, 155
    fasting 108–111, 149
Hyperinsulinaemia 52, 65, 117
Hyperinsulinism 23, 24, 100, 116
Hyperketonaemia 111, 112
Hyperlipidaemia 132, 133, 144, 145,
    154, 156, 157
Hypertension 134–136
Hypertriglyceridaemia 132, 136, 149–
    158
Hypoglucagonaemia 100
Hypoglycaemia 22, 23, 52, 78, 116,
    117
Hypoinsulinaemia 65, 96, 97, 114
Hypophysectomy 176
Immunoassay
    C-peptide 15, 21–23
    proinsulin 14
    synalbumin 185
Immunorective glucagon (IRG) 70, 71
    71
Immunoreactive insulin (IRI) 14, 16–
    18, 22, 23, 24
Incidence of diabetes in relatives 177–
    182, 183
Inheritance 52, 53, 177–182, 187, 189
Inhibition
    adenylcyclase 42
    adrenaline 42, 47, 49
    ATP concentration 46
    calcium uptake 40
    cyclic 3',5'-AMP phosphodiesterase
        41, 42
    diguanides 81
    gastric inhibitory polypeptide (GIP)
        49, 69, 72, 75, 78, 82
    glucagon secretion 95
    gluconeogenesis 100, 103
    glucose uptake 100, 109
    glycolysis 44
    insulin release 37, 40, 41, 49
    mannoheptulose 12, 37, 39, 45, 46,
        47, 51, 55
    nickel 40
    noreprinephrine 66
    nucleoside transport 36
    plasma triglyceride production 155
    proteolysis 5
    secretin 72
    somatastatin 95
    splanchnic glucose output 115

205

Initiators
 insulin release 51, 54
 insulin secretion 37
Insulin
 antagonism of plasma albumin
 173–176
 artefactual antagonists 174, 175
 artery, activity in 146, 147
 basal secretion 137
 big, big 16
 biosynthesis 1–14, 32, 33
 circulating antibodies 21–23
 conversion from proinsulin 8–10
 crystallization 11
 effectors of release 37–39
 glucose uptake, effect on 185, 186
 gut, effect on 80
 hepatic sensitivity 100–103
 immunoreactive insulin (IRI) 14
 initiators of release 51, 54
 insulin-releasing polypeptide (IRP)
 75
 insulin specific protease (ISP) 14,
 18, 19
 intravenous 174
 plasma 52–55, 56, 94, 172, 173,
 191,192
 release; see Release, insulin
 resistance; see Resistance, insulin
 secretion; see Secretion, insulin
 storage 32, 33, 39
 structure 2
 treatment 137, 147
Insulinoma 77, 78
International Atherosclerosis Project
 130
Intralipid 151
Intravenous glucose 48, 52, 53, 64,
 65, 128, 177, 188, 191
Intravenous insulin 176
Iodoacetamide 51
Iodoacetate 44, 45
Ischaemic vascular disease 127–131,
 194
Islets of Langerans 1, 4, 5, 6, 31,
 55
 ATP concentration 46
 cell tumours 18, 19
 glucagon secretion 70
 secretory products 14–24
 transplantation 112

Juvenile-onset diabetes (serum CPR)
 22, 23

Ketoacidosis 22, 111, 112, 149, 150,
 157
Ketogenesis, hepatic 112, 113
Ketosis 97, 148, 170
Kidney (glucose synthesis) 97, 104
Kinetics of insulin release 38, 39

Lactate 94, 99, 100, 105, 110, 114
Lactescent serum 149
Laennec's cirrhosis 117, 118
Lesions, atheromatous 140–144
Leucine 50
Linoleic acid 142
Lipaemia 152
Lipase 137
 lipoprotein 151,153, 154, 157, 158
Lipid metabolism (atherosclerosis)
 132, 133, 139, 140, 148–158
Lipodystrophy, congenital total 157,
 158
Lipoproteins 142, 144, 145, 148, 149,
 151
 lipase 151, 153, 154, 157, 158
Liver disease 116–119
Low density lipoprotein (LDL) 144
Lysine 11

Management of diabetes 113
Mannoheptulose 12, 37, 39, 45, 46,
 47, 51, 55
Mannose 47
Messenger RNA (mRNA) 6–8, 12,
 14
Metabolic clearance rate (MCR)
 C-peptide 21
 proinsulin 18
Metabolism
 C-peptide 21
 glucose 47, 48
 lipid 132, 133, 139, 140, 142, 148–
 158
 proinsulin 18–20
Methionine 7
Microangiopathy 125
Microfilaments 34, 36
Microtubules 34–36
Microvilli formation 33
Miller's technique 185
Molecular weight
 C-peptide 20
 glucagon 70
 glucagon-like   immunoreactivity
 (GLI) 71

precursors 6
proinsulin serum 16
synalbumin 181
tubulin 35
Monozygotic twins 52, 53, 64
Mortality in diabetics 125, 126, 129, 130, 133
Motolin 72, 75, 82
Muscle glucose utilisation 97, 103, 104, 112
smooth muscle cells 140, 141

Necrobiosis lipoidica diabeticorum 183
Nephrotic syndrome 155, 156
Nervous influences 63, 64, 66, 81, 82
Neurin 36, 37
Neurostenin 36, 37
Nicotinic acid 150
Norepinephrine 66

Obesity 52, 113–116, 172, 173, 182, 183, 192, 193
atherosclerosis 132, 133, 134, 155, 194
Offspring of diabetics 181
large babies 182
Oleic acid 142
Oral glucose 48, 49, 54, 64, 65, 97–99, 106, 108, 128, 172, 177, 188, 191
insulinoma 77
Oxford necropsy study of artery disease 127

Pancreatitis 156, 157
Parasympathetic nervous system 63, 66
Peptic ulcer 79
Pernicious anaemia 79
Phenformin 81, 129
Phosphodiesterase, cyclic 3′,5′-AMP 41–43, 50, 54, 55
Phospholipid 140, 143
Plasma
albumin, insulin antagonism of 173–176
cholesterol 145, 149
insulin 52–55, 56, 94, 172, 173, 191, 192
triglyceride 151–153, 155
Polyol pathway (sugar alcohols) 145, 146

Polyphagia 193
Post-heparin lipolytic activity (PHLA) 151–153, 157, 158
Potentiators of insulin secretion 37
Pre-diabetics 64, 65, 81, 173, 182, 191, 193
Pregnancy of diabetic mothers 182 190
Prevalence
atherosclerosis 127, 130
hypertension 135
hypertriglyceridaemia 149
Progranules 6–8, 10
Proinsulin
activation 4
amino acid secretion 13
circulating levels 16–18
conversion to insulin 8–10
metabolism 18–20
proinsulin-like component (PLC) 14, 16–18
RNA, messenger 6–8, 12, 14
serum 14–16
structure and properties 3,4
transport 32
Protein breakdown (starvation) 96
Proteolysis 3–5, 8, 10, 11, 32
Pyruvate 94, 99, 100, 105, 110

Radioactive labelling 6, 7, 8, 10
Radioimmunoassay
CCK-PZ 74
C-peptide 20
gastrin 67
glucagon-like immunoreactivity (GLI) 70, 77
glucagonomas 78
gut hormones 82
insulin secretion 81
plasma insulin concentration 52
secretin 73, 74
Rat diaphragm model 172, 173, 176, 180, 184, 185
Regulation of insulin biosynthesis 11–14
Release, insulin 172, 189
amino acids 49, 50
hormones 48–50
kinetics 38, 39
mechanism 33–37
pharmacological agents 50, 51
secretin 73, 74
Removal of plasma triglyceride 151–153

Renal failure 155
Resistance, insulin
  lipodystrophy 157
  uraemia 155
Risk factors (atherosclerosis) 130, 134
RNA, proinsulin messenger 6–8, 12, 14
Rough endoplasmic reticulum (RER) 32

Secondary   hypertriglyceridaemias 155, 156
Secretin 65, 69, 71, 72, 73, 74, 76, 82
Secretion granules 8–11
  properties 33
Secretion of C-peptide 21–23
Secretion, insulin 43, 52–56, 64, 65, 107, 178
  atherosclerosis 137, 138
  gastrin 74, 75
  gastrointestinal hormones 76
  glucagon 75
  potentiators 37
  stimulation by glucose 12, 36, 43–47
Secretory products, islet 14–24
Severity of diabetes (atherosclerosis) 131
Smooth muscle cells 140, 141, 143–147, 148
Somatostatin 95
Sorbitol 141, 146
Sphingomyelin 140
Splanchnic glucose output 97–100, 102, 104, 105, 109 110, 111, 112, 113–115
Stability of diabetics (C-peptide estimations) 23
Starvation
  diabetes 56
  normal man 95–97
Stenin 36, 37
Stomach 79
Streptozotocin 138
Sugar alcohols 145, 146
Sulphonylureas 42, 50, 80, 81, 156
Surgically-produced diabetes 138
Sympathetic nervous system 63, 66
Synalbumin negative 178–181
Synalbumin positive 178–181

Synthesis
  cholesterol 142, 143
  fatty acid 143
  triglyceride 150

Tecumseh prevalence studies 129, 130, 132, 134, 135
Tetracaine 40
Theophylline 41, 55
Tolbutamide 39, 76, 80, 131, 177, 187
Transduction of response to sugars 48
Transport, triglyceride 151–154
Treatment of diabetes 131
  atherosclerosis 131
  insulin 137, 147
Trichloracetic acid (TCA) 174, 175
Triglyceride 138, 143, 144, 148, 149–155
Trypsin 6, 8, 16
Tubulin 35, 36
Tumours
  hormone-secretory 77–79, 82, 83
  islet cell 18, 19
Tyrosyllated C-peptide 20

Ulcer, peptic 79
University Group Diabetes Program (UGDP) 131, 132, 138
Uraemia 155

Vagus nerve 66, 82, 184, 188
Vasoactive intestinal peptide (VIP) 69, 72, 75, 78, 82
Verner Morrison syndrome 78
Very low density lipoprotein 144, 145, 150
Vinblastine 35, 36, 37
Vincristine 35
Viral hepatitis 116, 117
Visking 175
Vitiligo 183

Watery   diarrhoea   hypokalaemic achlorhydric syndrome (WDHA) 78
Weight loss 155

Xanthomata diabeticorum 156, 157

Zinc binding 11
Zollinger–Ellison tumours 78, 79